DRIVEN

The Race to Create the Autonomous Car

ALEX DAVIES

SIMON & SCHUSTER

New York   London   Toronto   Sydney   New Delhi

Simon & Schuster
1230 Avenue of the Americas
New York, NY 10020

First Simon & Schuster hardcover edition January 2021

SIMON & SCHUSTER and colophon are registered trademarks
of Simon & Schuster, Inc.

For information about special discounts for bulk purchases,
please contact Simon & Schuster Special Sales
at 1-866-506-1949 or business@simonandschuster.com.

The Simon & Schuster Speakers Bureau can bring authors to your
live event. For more information, or to book an event, contact the
Simon & Schuster Speakers Bureau at 1-866-248-3049
or visit our website at www.simonspeakers.com.

Interior design by Paul Dippolito

Manufactured in the United States of America

1   3   5   7   9   10   8   6   4   2

Library of Congress Cataloging-in-Publication Data
Names: Davies, Alex, author.
Title: Driven : the race to create the autonomous car / Alex Davies.
Description: First Simon & Schuster hardcover edition. | New York : Simon &
Schuster, 2020. | Includes bibliographical references and index.
Identifiers: LCCN 2019042560 | ISBN 9781501199431 (hardcover) | ISBN
9781501199455 (trade paperback) | ISBN 9781501199462 (ebook)
Subjects: LCSH: Automated vehicles--History. | Automobile industry and
trade--Technological innovations--United States. | Competition.
Classification: LCC TL152.8 .D38 2020 | DDC 629.22209--dc23
LC record available at https://lccn.loc.gov/2019042560

ISBN 978-1-5011-9943-1
ISBN 978-1-5011-9946-2 (ebook)

For my grandfather, Patrice Lestelle—
a great lover of books and a terrible driver of cars.

Tell me of your country,
Your people, and your city, so our ships,
Steered by their own good sense, may take you there.
Phaeacians have no need of men at helm
Nor rudders, as in other ships. Our boats
Intuit what is in the minds of men
And know all human towns and fertile fields.
They rush at full tilt, right across the gulf
Of salty sea, concealed in mist and clouds.
They have no fear of damages or loss.

—Homer's *The Odyssey*

Contents

DRIVEN

Prologue: *Waymo v. Uber*

A LITTLE AFTER NINE IN THE MORNING OF A COOL FRIDAY IN April 2017, Anthony Levandowski sat down where so many of his colleagues and friends had predicted he would land himself: in a conference room surrounded by lawyers, being grilled about his starring role in the first great battle of a world he had helped create.

If the blinding morning sun hadn't been coming through the window of the twenty-second-floor office in downtown San Francisco, Levandowski would have been able to see the Bay Bridge. Every day, 260,000 vehicles used the 8.4-mile span to cross the bay that divided the city from Oakland, Berkeley, and the rest of its East Bay neighbors. By six in the morning, the mass of cars, trucks, vans, and motorcycles waiting to pay the ever increasing toll and funnel onto the crossing created a mile-long parking lot. On days when someone crashed on the bridge, the resulting extra congestion could cripple the region's road network. Like eighteenth-century urbanites emptying chamber pots from upper story windows, it was a quotidian sort of insanity, excused by entrenchment and a lack of better options.

Attorney David Perlson, of the white shoe law firm Quinn Emanuel Urquhart & Sullivan, began the deposition. "Where do you work currently?"

"I work at Uber," Levandowski said.

Six feet six inches tall and slim, with a head of dark hair that was starting to recede, Levandowski wore a blue suit for the occasion, no tie. Apart from the black sneakers, it was a rare change from the standard

Silicon Valley engineer look he embraced: jeans and whatever T-shirt was on top of the dresser drawer that morning.

"Okay," Perlson said. "And what's your position there?"

"I'm vice president of engineering."

"What are your responsibilities as vice president of engineering?"

Here, at the direction of his lawyer, Levandowski read from a piece of paper on the table in front of him.

"On the advice and direction of my counsel, I respectfully decline to answer," Levandowski said. "And I assert the rights guaranteed to me under the Fifth Amendment of the Constitution of the United States."

"How long have you worked at Uber?"

"On the advice and direction of my counsel, I respectfully decline to answer. And I assert the rights guaranteed to me under the Fifth Amendment of the Constitution of the United States."

Over the following six hours, Levandowski declined to answer one question after another, questions that in their one-sidedness built a damning narrative.

"When you worked at Google, you received tens of millions of dollars in compensation from Google, is that accurate?"

"You and Uber discussed how you would form a new company while you were employed by Google?"

"You and Uber discussed that your new company would eventually be acquired by Uber while you were still employed at Google?"

"That new company eventually became Otto, correct?"

"While you were still employed by Google, you recruited engineers to join your new company so that your new company could replicate Google's Lidar technology, correct?"

"You took over fourteen thousand confidential files from Google prior to your departure from Google, correct?"

"You took the fourteen thousand documents from Google so that you could get—so that you could more quickly replicate Google's technology at Otto, correct?"

"Mr. Levandowski, your use of the fourteen thousand confidential documents you took from Google allowed you to sell Otto to Uber for over $680 million in just a few months?"

Again and again and again, Levandowski gave his carefully scripted nonanswer, citing his Fifth Amendment rights.

Officially speaking, Levandowski was just one of many witnesses being deposed in the run-up to *Waymo v. Uber*, a legal brawl between two corporate giants. Waymo had started life as a Google project called Chauffeur, and was now its own company under the umbrella of Google's parent company, Alphabet. Uber was the enormously valuable ridehailing company that had thrown the world of urban transportation into chaos since its founding in 2009. Both were racing to create and deploy cars that could drive themselves.

Their fight centered on the thirty-seven-year-old Levandowski, who had spent nine years at Google before moving to Uber. In Waymo's telling, on December 14, 2015, Levandowski downloaded more than fourteen thousand technical files from its servers onto his laptop, many of them describing the inner workings of its all-important Lidar laser vision system. He connected an external hard drive into the computer for eight hours, then installed a new operating system to wipe away evidence of the downloads. He quit six weeks later and founded Otto, a company dedicated to developing self-driving trucks. After a few months, Uber acquired Otto for a reported $680 million—an astounding figure for such a young company—and put Levandowski in charge of its own autonomous driving project.

Under Levandowski's direction, Waymo alleged, Uber's engineers used those files to accelerate their technical progress and play catchup, having started their research only in 2015, six years after Google. That, Waymo insinuated, was why Uber had been able to send robotic trucks along the highways of Colorado and Nevada, how it was using robotic cars to move people around Pittsburgh. Those vehicles still had people behind the wheel, but it was only a matter of time—time better counted in months than years—before the flesh-and-blood backups were no longer necessary.

Uber said that nothing Levandowski may have taken made its way into its work.

If Waymo's phalanx of lawyers convinced the jury that Uber had cheated to get ahead, Uber could be forced to put its autonomous driving efforts on ice, or maybe the scrap heap. And that wouldn't be just a hit to the balance sheet. It would be an existential crisis. Driverless cars would be safer and cheaper than human-driven ones, and any service that provided them would dominate the market, said Uber CEO Travis Kalanick. "In order for Uber to exist in the future, we will likely need to be a leader in the AV, autonomous vehicle, space."

Kalanick was right. Robots will drive the future. By the start of the *Waymo v. Uber* trial in February 2018, fleets of autonomous vehicles were roaming the streets of Silicon Valley, San Francisco, Pittsburgh, Phoenix, Detroit, Boston, Munich, and Singapore—to name a few. Tesla, Cadillac, BMW, Audi, Mercedes-Benz, Nissan, and other automakers were selling cars that could pilot themselves on the highway. Along with Google and Uber, Ford, General Motors, and others were working on fully driverless cars that wouldn't need steering wheels or pedals. Dozens of companies, from the world's largest corporations to the smallest startups, were crowding into a technology whose upside flirted with utopianism. The average American worker spent nearly an hour driving to and from work every day; driverless technology would turn that chore into free time. Robots that never get drunk, tired, angry, or distracted promised to drastically reduce crashes, more than 90 percent of which result from human error. Those crashes kill about forty thousand Americans every year. Globally, the annual death toll is well over a million.

Uber and Waymo executives sang sweet songs about ending road deaths, but they weren't in court fighting over who got to save more lives. They went to war because each wanted to claim a dominant share of a market predicted to be worth $42 billion in 2025 and $77 billion in 2035, when 12 million new robo-cars would hit the road annually. By 2050, autonomous driving tech could add $7 trillion to the world's economy, all of it for the taking by anyone who could make it safer to get around, cheaper to move goods, and way more relaxing to be stuck in traffic.

That was the near term. The advent of the personal car shaped the world's cities, suburbs, and rural areas over the past century. It created cultures. It inspired art; it was art. It helped create and define the middle class. Questions remained about how autonomous cars would be tested, certified, insured, and operated. But these were details. The shift away from human driving promised to be as influential as the car itself, if not more so. It offered the opportunity to remake cities, to correct the mistakes of the past.

Driverless cars would be shared, and they'd be cheaper than today's taxis or Ubers. They wouldn't need to take up precious urban space for parking, instead driving themselves to lots in less dense areas. They'd run on electricity instead of gasoline, reducing pollution and helping balance the power grid. They'd boost productivity. Many more effects were hard to anticipate. Just as the smartphone begat an app ecosystem, including a ridehailing market dominated by Uber, robotic driving could create entirely new industries.

Critics and skeptics feared the tech would encourage suburban sprawl, since people wouldn't mind long commutes if they could work, sleep, or relax on the road. The promise of smarter cars could sap officials' interest in funding reliable, equitable public transit. Self-driving cars could ruin the fun for people who like being behind the wheel, and multiply cybersecurity risks for everyone, giving malevolent hackers a juicy new target. And they were poised to eliminate, over the coming decades, the jobs of the 4 million Americans who made a living by driving.

The reality, though was inescapable. The age of autonomous vehicles was coming, and—like sails, steam, combustion engines, and the physics of flight—the technologies propelling it along would turn the world on its head.

By the time Anthony Levandowski stepped into the conference room for his deposition, he had seen his reputation, and possibly his career, transformed into a smoking ruin. The judge called Waymo's account "one of the strongest records I've seen for a long time of anybody doing something that bad." In May, he took the unusual step of recommending that the Department of Justice investigate criminal charges. Uber fired Levandowski a week later.

Pleading the Fifth—Levandowski would repeat those two sentences 387 times in that day's deposition—may have protected him legally, but it also meant that at the moment his critics were loudest, no one spoke in his defense. No one would say that he had been there at the beginning of Google's self-driving research, that he had done more than perhaps anyone else to bring this technological revolution to the brink of reality. But at the end of that Friday in April, with twenty minutes left in a six-hour deposition, Uber's lawyer asked him how he had heard about something called the DARPA Grand Challenge, while Levandowski was a graduate student at the University of California, Berkeley. Since the topic had little to do with the facts of the case, his lawyer let him respond. The answer took him back to a 2003 conversation with his mother.

"My mom knew how much I loved robots and that I loved making things. And she gave me a call when she found out about this competition sponsored by the Defense Department," Levandowski said. "When I saw it, I couldn't resist.

"It was a race from LA to Vegas, across the desert," he went on. "The goal was to release a vehicle into the world on its own without any remote control or assistance" and have it go from start to finish, all on its own. His entry, he said, was a motorcycle called Ghostrider. "I built a substantial portion of it myself, but I also created a team to help me," Levandowski said. "It was, frankly, a pretty crazy idea." The self-riding motorcycle didn't reach the other side of the desert, but it did make its way across the country when America's great museum came calling. "I donated it to the Smithsonian, where it is today."

A few questions later, the lawyers were back to the more recent past and Levandowski was back to the Fifth Amendment, back to a defensive silence. But his contribution to the annals of technological history didn't end with landing a robotic motorcycle in the Smithsonian. That was simply where it began.

—1—

The Grandma Test

THE SIGNING OF THE FLOYD D. SPENCE NATIONAL DEFENSE AU-thorization Act for Fiscal Year 2001 didn't warrant a Rose Garden ceremony, a bouquet of microphones, or a write-up in the next day's *Washington Post*. The 515-page document was routine legislation, setting the budget for the American military: which weapons it would build, how much veterans would pay for prescription medications, which rusting artifacts would be transferred to museums.

President Bill Clinton, on his way out of office, had his quibbles with the bill, sent to his desk by a Republican majority Congress. But he deemed it fine in the balance, and necessary to the nation's security. In a statement, he praised the bits he liked: the increased housing allowances for military personnel, the authorized cleanup of a former uranium mill in Utah, funding for the next-generation F-35 fighter jet. The president had nothing to say about Section 220, which read:

It shall be a goal of the Armed Forces to achieve the fielding of unmanned, remotely controlled technology such that—

(1) by 2010, one-third of the aircraft in the operational deep strike force aircraft fleet are unmanned; and

(2) by 2015, one-third of the operational ground combat vehicles are unmanned.

For the staffers and lobbyists who wrote the bills on which legislators stamped their names, this sort of mandate was a common tool for getting things done, or at least securing the funding to try. Section 220, along with most of the bill, came from the office of John Warner, the Virginia senator who helmed the Armed Services Committee. Warner had enlisted in the navy as a seventeen-year-old in 1944, joined the Marines during the Korean War, and served as Richard Nixon's secretary of the navy (marrying and divorcing Elizabeth Taylor along the way). By 2000, he had been a senator for more than two decades, and saw the role robotics could play in the future of warfare. The Predator drone had entered service over the Balkans in 1995, letting American pilots "fly" over dangerous territory without risking their lives.

Warner wanted the US to rely far more on such tools, even if the military wasn't raring to make such a drastic change. "We wanted to move swifter, more forward leaning," Warner said. "The Pentagon wanted to follow its usual, more conservative track." A mandate, he figured, might change that attitude.

With the Predator already in service, the first part of the Section 220 mandate was just a matter of multiplying that success, applying it to more aircraft and pumping out more drones. Unmanned ground vehicles were less developed, but at the time, the advent of trucks and tanks that could drive without a person on board seemed plausible, maybe even imminent. Computers could weave a fighter jet through the air, launch a ballistic missile from a submarine, or destroy a target from a hemisphere away. Researchers in the United States, Asia, and Europe had demonstrated vehicles that could drive themselves in restricted conditions. Les Brownlee, the staff director for the Armed Services Committee, who helped Warner craft the bill, thought that with a fifteen-year window, making robotic vehicles a major presence within the armed forces was doable. And he knew America's scientists wouldn't deliver without a push. "We certainly wanted to challenge them," he said.

It made perfect sense, except to people who happened to know anything about unmanned technology. The sky is virtually empty, so you don't need much more than a good understanding of aerodynamics to fly a drone like the Predator. Driving demands the ability to find and stick to flat, or at least even, ground, and to contend with rain, snow, and fog that can blind computer vision systems, but that aircraft can fly above. It requires not just avoiding all the things gravity keeps out of the sky—trees, rocks, buildings, people, other vehicles—but understanding what they are, how they're likely to act, and how one's own movement affects others' plans. Driving might be the most complicated task humans undertake on a regular basis, even if they don't realize it.

Moreover, while Warner's mandate called for "unmanned" vehicles, the Predator drone's remote control setup was a nonstarter on the ground. Because flight requires relatively few split-second decisions, latency—the delay between a pilot sending a command and seeing it executed—is more pesky than problematic. When navigating the crowded ground at "tactically relevant speeds" of fifteen or twenty miles per hour, it's devastating. (Think of elderly drivers with slowed reaction times.) Remote operation might demand a one-person, one-robot paradigm that left major benefits on the table, like reduced need for manpower. Like regular driving, it requires one's full attention. If a soldier in the field wanted to remotely control a scout vehicle, she might need someone to stand watch for her—doubling instead of slashing staff requirements.

The challenge Bill Clinton signed into law wasn't to connect a human to a remote vehicle. It was to teach a car to do everything a human can do. Anyone familiar with this technology who read Warner's mandate knew that America's unmanned ground vehicles would have to be autonomous. Warner may not have recognized the difficulty of the challenge he put into law. But he knew he wanted it done, and he knew who might be able to do it.

On June 18, 2001, Tony Tether walked into an office on the ninth floor of 3701 Fairfax Drive in Arlington, Virginia. The Stanford-trained engineer had spent plenty of time in this room in the 1980s, always on the side of the desk closer to the door. This time, though, he sat behind the desk, as President George W. Bush's newly confirmed choice to run the Defense Advanced Research Projects Agency.

Better known by its acronym, DARPA was born into the Pentagon's sprawling organizational tree in February 1958, as a response to the Soviet Union's launch of *Sputnik 1*. That small, pinging satellite, which circled the planet every ninety-eight minutes and was visible from Earth, shook Americans and their government. Dwight Eisenhower wanted an agency dedicated to ensuring the United States would never again be surprised by a technological advance, an agency that stood apart from the army, navy, air force, and Marines. The small outfit was first called the Advanced Research Projects Agency, or ARPA. As the military nature of its mission morphed, the "D" for "Defense" was added in 1972, removed in 1993, and put back in 1996.

Like its name, DARPA had an unsettled, roving history. It started as America's de facto space agency, then lost the role to NASA when the civilian agency was created a few months later. Without much of a mission statement or specific goals, the runt of the Pentagon spent its first decade focused on missile defense at home and counterinsurgency in Southeast Asia, researching new ideas and funding scientists with promising pitches. Those efforts produced what would become DARPA's standard mix of results: nonstarters, embarrassing failures, and a heavy helping of projects whose impacts spread beyond whatever anyone had imagined or intended.

First tasked with the devilish problem of defending the United States against nuclear attack, DARPA explored ideas for a particle beam gun that could shoot down an incoming ICBM. That went nowhere. The agency fared better when it worked on the ability to detect Soviet nuclear weapons testing. Along with developing technology to spot such testing in outer space, DARPA installed seismographs all over the planet and funded research to identify tremors as natural—e.g., earthquakes—or the result of an underground nuclear test. That laid the groundwork for the 1963 Limited Nuclear Test Ban Treaty, by giving the US confidence in its ability to spot Soviet cheating. Meanwhile, DARPA's support of seismographic research proved invaluable to the scientists who presented the theory of plate tectonics.

DARPA's greatest success started in 1961, when Joseph Carl Robnett Licklider joined the agency to do some behavioral science work and improve the military's ability to counter conventional and nuclear weapons in times of crisis. Licklider was a psychologist with a deep interest in the budding field of computing. He focused his energy on the command-and-control assignment, which he saw as one of many potential applications for his grand vision: a network of computers that did more than arithmetic. He funded research at places like MIT, Stanford, and the defense-oriented think tank RAND, aiming to connect a few computers in the same room. In 1965, one of Licklider's successors, Robert Taylor, decided to pursue the idea on a grander scale. In a fifteen-minute meeting, he squeezed a million dollars out of his boss and used it to create the ARPANET—the network that became the internet.

Less eulogized is DARPA's work in Southeast Asia. In May 1961, the agency launched Project AGILE, a counterinsurgency program proposed by William Godel. An intelligence operative and one of the agency's first employees, Godel cranked out innovative, often absurd ideas for helping embattled Vietnamese president Ngo Dinh Diem fight the Communists coming from the north. DARPA experimented with portable flamethrowers, mines made to look like rocks, and a near-silent "swamp boat" that could carry thirty men through water just three inches deep. But Godel was especially interested in destroying the crops and jungle foliage that fed the Viet Cong and let them covertly move supplies and launch ambushes. DARPA funded the development of a range of chemicals, millions of gallons of which American C-123 cargo planes would pour over South Vietnam. The best known of these was called Agent Orange. It ravaged the land and left behind a trail of cancers and birth defects that devastated Americans and Vietnamese alike. As antiwar sentiment built up at home, DARPA was moved from its original office in the Pentagon to the Fairfax Avenue building in Arlington—a physical manifestation of its bruised reputation.

These diverse efforts were all born of DARPA's defining trait: flexibility. The agency worked nothing like the rest of the military. It usually employed no more than a few hundred people and was largely unbound by the bureaucracy that dictated life in most of the government. The director had the office on the top floor, but the direction came from the program managers who made up more than half the head count. These were physicists, chemists, biologists, and engineers, academics and industrialists, civilians and service members. Their job was to come up with potential solutions to stubborn problems they encountered, a new kind of communication device or armor or navigation system. They pitched the director on the program they wanted to run and, if approved, found and funded the companies or universities or whomevers who could make their ideas real. Program managers often lasted just a few years. Few went more than five. DARPA favored constant turnover, prioritizing new thinking over institutional memories, especially of failures. When a project worked, DARPA handed it off to the military or private sector for commercialization, and went looking for the next wild venture.

By the time Tony Tether first came to DARPA in the eighties, this approach—hunting down innovative leaps to solve real problems, dodging bureaucracy all the while—had produced or laid the groundwork for the stealthy F-117A fighter jet and B-2 bomber, the M-16 rifle, the Predator drone, and GPS. Then forty years old, Tether had the look and CV of a defense industry lifer. He wore his hair slicked down and seemed to have stopped buying new glasses around the time he got his PhD in electrical engineering, in 1969. Tether spent four years as the head of DARPA's Strategic Technology Office, doing work that remains classified. When the DARPA director job opened up in 1985, Tether went for it and lost. He returned to the private sector, where he stayed until Defense Secretary Donald Rumsfeld brought him in for an interview.

Along with his engineering background, Tether's love for science fiction made him a good fit to run DARPA. As a kid, he had listened to *Sputnik* beeping overhead on his ham radio. He was enamored of novels like Robert Heinlein's *The Moon Is a Harsh Mistress*, where humans colonize the Moon, then start a civil war with those who remain on Earth. "I believe strongly that the best DARPA project managers must have inside them the desire to be a science fiction writer," he said. H. G. Wells, he thought, would have been a fantastic employee. But by the time Tether sank into the director's chair in June 2001 and added a few personal touches—he didn't bother swapping out the old furniture—it hardly mattered whether his deputies had read any sci-fi, let alone written their own. America's great bogeyman, the Soviet Union, was long dead, and with it had gone the agency's motivating force. The 1990s had been about the peace dividend, not defense spending. Through the first summer of Tether's tenure, Americans weren't watching for an invasion or nuclear attack. The US was the world's lone superpower, and needed its mighty military only to swat at the occasional militant group in Africa or the Middle East. "DARPA had become a backwater," Tether said.

A few months later, on a sunny Tuesday morning in September, Tether's secretary pulled him out of a conference room and directed his gaze out the window, to the east. Black smoke was filling the sky over the Pentagon, DARPA's former home. Soon America was back at war. In Washington, defense once again took center stage, and the money flowed: From 2001 to 2005, DARPA's annual budget increased 50 percent, to $3 billion.

Right away, Tether diagnosed the attacks as resulting from a failure of intelligence. He wagered the clues were all there, just not in one place, where any one person or agency could put them all together. Within months, he launched an intelligence gathering project pitched to him by John Poindexter, then senior vice president of SYNTEK Technologies, as "A Manhattan Project for Combatting Terrorism." Poindexter was best known for his central role in the coverup of the Reagan-era Iran-Contra Affair, but Tether was willing to overlook his shady history.

He thought he was the right man to run a project they called Total Information Awareness. But before long, September 11 led to military questions that weighed more heavily on the public's mind than ferreting out terrorists.

In Afghanistan and Iraq, American men and women in uniform met a vicious antagonist: insurgencies using roadside bombs to kill and dismember the troops traveling local roads. As the hopes for a quick and glorious romp through the Middle East soured, Tether kept thinking about John Warner's unmanned vehicle mandate, and what DARPA could do to fulfill it.

The dream of a vehicle that drives itself dates back to the early days of the automobile, as people abandoned sentient horses for machines that punished any lapse in attention. In 1926, the *Milwaukee Sentinel* announced that a driverless "phantom auto" would tour the city, controlled by radio waves sent from the (human-driven) car behind it.

The idea went national with Futurama, General Motors' exhibit at the 1939 New York World's Fair. Millions of Americans braved hours-long lines for the chance to sit in the navy-blue mohair armchairs that would take them on a seventeen-minute tour of a "wonderworld of 1960." During the tour, when they weren't too busy necking with their sweethearts, they ogled massive dioramas of a national highway system that eliminated crashes and congestion, where radio control systems kept everyone in each of the fourteen lanes going a set speed and staying a safe distance apart. At the height of its power at the time, GM kept at the idea. A promotional video for its 1956 Firebird II concept car explained that "the driver might just push a button, and the car would literally drive itself" by picking up electronic signals from the highway. The automaker teamed up with RCA to build a test track in Princeton, New Jersey, but soon abandoned it as impractical at scale. In the 1960s and '70s, researchers at Ohio State; University of California, Berkeley; and in Japan and Germany did similar work.

All these concepts, though, were limited in scope to the easiest part of the driving problem, cruising on the highway. With the cars pointing in the same direction, all you needed was a way to keep them in their lanes and away from one another. Given the right mix of infrastructure and in-car tech, the problem seemed tractable, if hard to implement at a national scale. No one seriously considered making a car that could negotiate a more complex environment, with intersections, traffic signals, and pedestrians. Even in America's most enthusiastic portrayal of the future, the problem went untouched: George Jetson did his own driving.

That's because the kind of technology that might be able to mimic the human driver—who surveys his surroundings, analyzes their elements, predicts how the scene will evolve over time, and moves accordingly— just didn't exist. Not yet, anyway. While GM was laying underground cables, researchers in Palo Alto, California, were pioneering a field they called "artificial intelligence."

The first robot that could move around and "think" about its actions was Shakey, built by the Artificial Intelligence Center at Stanford Research Institute (which has since split from the university and changed its name to SRI International). But the first machine widely recognized as an autonomous *vehicle* came from nearby Stanford University's Artificial Intelligence Laboratory, or SAIL. The Stanford Cart had been built in 1961, part of research into how well a human on Earth might be able to control a rover on the Moon. This cart, which looked like a card table riding a quartet of bicycle wheels, spent the next twenty years being passed from one researcher to another, each using the platform for his own application. By the time the Austrian-born computer scientist Hans Moravec adopted it in the early 1970s, it could use a camera to follow a wide white line painted on the ground, in very specific conditions, at not quite 1 mph.

Moravec wanted to make the computer do more, and found his solution by watching some lizards he had caught and kept in a terrarium. Before pouncing on a fly, the lizards would fix one eye on their prey, then sway their head from side to side. Perhaps, Moravec thought, a computer could calculate the distance of objects it saw the same way. So he put the camera on a slider and programmed it to move from one side to the other, taking photos along the way. With this spin on stereo vision, the cart's computer (which took up most of a nearby room) would pick out spots of high contrast, the things most likely to be objects. By comparing their positions in the sequence of photos, it could fix the location of each in space. Moravec would let the cart loose in a large room or an outdoor space strewn with chairs, trees, and cardboard icosahedrons (imagine overgrown twenty-sided dice). It navigated slowly, pausing for ten to fifteen minutes between one-meter dashes. It bumped into things. But it navigated, covering one hundred feet in five hours. "Similar humble experiments in early vertebrates eventually resulted in human beings," Moravec wrote in his 1980 dissertation.

After completing his PhD, Moravec moved to Pittsburgh to join the newly created Carnegie Mellon University Robotics Institute, where he and his colleagues fostered that evolution. In 1984, their efforts in the budding field of computer vision—the ability for a machine to see and understand its surroundings—graduated to honest to goodness vehicles with the Navigational Laboratory program. NavLab 1 was a blue Chevy panel van that carried around an extra four-cylinder engine just to generate the power to run its onboard supercomputer, camera, laser scanner, and radar. The van did most of its early testing at speeds around 1 mph and couldn't do much more than follow the road and spot obstacles ahead. But it was maybe the first robot roomy enough for its makers to work inside it as it moved, making it a viable transportation option. As computers got faster and sensors improved, the CMU team produced a series of vehicles with increasingly humanlike abilities.

When they got to NavLab 5, they decided it was time for a road trip. Researchers Dean Pomerleau and Todd Jochem had developed a program called the Rapidly Adapting Lateral Position Handler (RALPH), which used a camera to look for lane markings, road edges, and discoloration from dripped oil to find the center of the road and stay there.

Eager to see how RALPH, installed in a gray Pontiac Trans Sport mini-van, would fare on highways with different sorts of road markings and types, in the summer of 1995 Pomerleau and Jochem launched a cross-country drive they called "No Hands Across America." (Since their focus was on the road vision system, they worked the gas and brakes.) Over nine days, listening to *Star Trek* books on tape to pass the time, the computer scientists covered the 2,849 miles to San Diego, letting the car do nearly all the steering.

America wasn't the only country with robots hitting the road. Even before Carnegie Mellon developed its NavLab vehicles, the Tsukuba Mechanical Engineering Lab in Japan created a vehicle that drove itself at 20 mph by using cameras to pick out lane lines. In the 1980s, Germany's Daimler launched a program it called Prometheus (that's "program for a European traffic of highest efficiency and unprecedented safety" in German). Collaborating with computer vision pioneer Ernst Dickmanns, Mercedes-Benz's parent company built a series of vehicles that drove themselves in simple (e.g., highway) settings at various speeds, culminating in October 1994, when an S-Class sedan with video cameras drove itself six hundred miles on a multilane motorway, even changing lanes and overtaking other cars. In an offshoot of the Prometheus program, Italian researchers made a car that drove them more than seven hundred miles.

Apart from GM's early toe dip, the American auto industry was neither interested nor involved in such projects. The funding for the research being done at Carnegie Mellon and elsewhere in the country came largely from the Department of Defense—and from DARPA in particular. The Pentagon's Skunk Works arm had long seen robotics as a vital capability for the future of warfare, and exactly the sort of long-view work it was designed to tackle. It had funded the scientists who built Shakey and some of Moravec's work with the Stanford Cart. In the early 1980s, DARPA started an autonomous land vehicle program, using an eight-wheeled ATV, which by 1987 could follow a curving road made up of different kinds of pavement. Through the 1990s,

DARPA collaborated with the army and sponsored research by various defense contractors and universities, focusing on making vehicles that could run scouting missions.

In total, decades of scattered and sporadic work by a variety of parties—chiefly government-funded academics—had produced a solid foundation for the kind of military force John Warner demanded. It wasn't much more than that, though. The elite robots of the early 2000s were slow, expensive, unreliable, and not a sure bet to beat the old Stanford Cart in a race. The idea looked like the eternal research project, one built of steady, incremental advances that would produce a steady stream of PhD dissertations.

Now Congress had told DARPA it had less than fifteen years to turn these raw robotic recruits into a nimble, hardy, and ubiquitous fighting force. Scott Fish, a program manager running one of the agency's robotics efforts, remembers the word coming down. "That's a hell of a challenge," he thought.

As he pondered how to meet Warner's mandate, Tony Tether thought of another line from Congress, this one tucked into the *National Defense Authorization Act for Fiscal Year 2000*:

> The Secretary of Defense, acting through the Director of the Defense Advanced Research Projects Agency, may carry out a program to award cash prizes in recognition of outstanding achievements in basic, advanced, and applied research, technology development, and prototype development that have the potential for application to the performance of the military missions of the Department of Defense.

Tether didn't have to let out yet another series of contracts and hope America's academics and defense contractors turned up a miracle. He could join a storied history of innovation triggered by competition.

He would create a twenty-first-century version of the Orteig Prize, the $25,000 reward offered by a New York hotelier that Charles Lindbergh won by flying solo across the Atlantic.

The idea of a contest had been kicking around DARPA before Tether came on as director. That snippet of law was the product of Rick Dunn, the agency's general counsel, who worked to enhance the agency's prized flexibility. He convinced legislators to give DARPA the right to enter into partnerships with commercial companies and to dodge civil service laws so it could better recruit scientists and engineers. The ability to award prizes instead of contracts was just one more element in what Dunn called "an ecosystem of doing business in an innovative way."

The prize authority didn't make much sense for most DARPA projects. If the goal was to develop a stealth plane or new sort of rifle, the standard sort of contract with one or a few defense industry contractors was the right bet. But Tether recognized a particular swirl of factors that made an unmanned ground vehicle the ideal application for a contest open to the public.

First, the impetus was there, in the form of the Warner mandate and in the rapidly developing conflicts in the Middle East that threatened to draw in more and more American soldiers. Second, the technological foundation had been laid. Cars and computers were readily available. Anyone with the right phone number and some cash could get the servers, cameras, radars, or laser range finders a vehicle might need. Third, the intellectual foundation was there, put down and published by researchers at Carnegie Mellon, Stanford, and elsewhere.

Tether believed the trick at this point was more integration than invention, finding just the right way to mix together all the existing pieces. His quest was for the secret sauce, and he couldn't have too many cooks. He wanted to bring in brilliant folks who had nothing to do with the Department of Defense or DARPA. Tether just needed to find them and lure them into the kitchen.

Moreover, the idea of a contest appealed to Tether in a way no contract ever could. He was a scientist, but he was also a salesman. Decades earlier, between finishing college and starting his graduate studies at Stanford, the newly married Tether had gone door-to-door hawking home cleaning products for the Fuller Brush Company. He was good at it, learning to capture people's attention, to wriggle his way inside and sell them on a vision, even one based around a mop.

A few weeks into his tenure at DARPA, when Vice President Dick Cheney came by for a briefing on the agency's current projects, Tether planned his lineup like a manager going into the World Series. Ever the salesman, the director started with DARPA's most exciting work. Figuring Cheney wouldn't stick around too long, Tether saved his less compelling employees for last. From start to finish, he thought about the story of each program. "I picked them if I thought, 'Wow! These are really important, and not only are they important, but they can be briefed in a way to show that they're important.'" Tether knew Cheney from the VP's time in the House of Representatives, and made sure the presentations focused on pictures, not words. By Tether's account, Cheney—joined by Secretary of Defense Donald Rumsfeld—loved the briefing, and left Fairfax Avenue as a DARPA ally.

In the same vein, Tether changed the tone of the DARPA Systems and Technology Symposium. The agency held this meeting, better known as DARPATech, every two years or so to brief anyone interested on what it was up to. It had long been a sedate, serious affair, held in a city like Dallas or Kansas City. Tether moved it to a Marriott in Anaheim, California, across the street from Disneyland. The location was more than subtext.

"Disneyland is a land of dreams and fantasy becoming reality," Tether said in his opening remarks at the DARPATech held in August 2002. "That is what DARPA does—and does well." He had made each branch of the agency pick a theme to fit the occasion. The Information Exploitation Office got the Sorcerer's Apprentice. The Tactical Technology Office was Frontier World. The Microsystems Technology Office took It's a Small World. "We were there to tell people what we were doing, and to get people to come with ideas," Tether said later. "I knew it was a show."

Before ceding the podium to a succession of deputies who would go over everything from a "self-healing minefield" to "fiber lasers," Tether made one more announcement. In addition to all its standard programs (if you call monkey-based mind control experiments standard), DARPA was planning a race for fully autonomous vehicles. It would take place at the next DARPATech conference, sometime in 2004, and run from Los Angeles to Las Vegas. Whoever finished in first place would take home a million dollars.

Tether called it the DARPA Grand Challenge.

Tether announced the race before sorting out the details. He hadn't scheduled the next DARPATech conference yet. He figured the cars could run on the I-15 freeway, but wasn't sure how, or if, he could take over a major interstate. He didn't know who might want to build an autonomous vehicle, or how long they would need. He had no idea how to put on a race, let alone one for robots. "We honestly did not know what we were doing," he said. But he knew he wanted something that would pass what he called "the grandma test." It had to be so straightforward that anyone could watch and know who won.

To sort out the realities, the director called on Jose Negron. The air force colonel had been assigned to DARPA in September 2001, arriving just days before the terrorist attacks that supercharged the agency's role in America's arsenal. Now his job included turning Tether's idea into a plan.

Negron's first move was informing his boss there was no way DARPA had the clout to shut down an interstate, let alone one that ran through Los Angeles. They wouldn't be able to send robots charging into downtown Las Vegas, either. Negron suggested starting in Barstow, an old railroad town about a hundred miles northeast of LA. The race would go mostly off-road, through the Mojave Desert. It would finish in Primm, a sad sack gambling town just over the Nevada border, frequented by Californians too eager to drive the next forty miles to Vegas.

Anyway, Negron thought, the Mojave was a more appropriate testing ground for military vehicles than a freeway. The land between Barstow and Primm—about a hundred miles as the crow flies—included fire roads, dirt paths, and the occasional stretch of pavement. It featured a mix of flat, open terrain and steep hills with narrow paths. In other words, it approximated many of the driving environments of the Middle East. If a vehicle could navigate the Mojave, it could handle Afghanistan and Iraq.

Running through the desert would also let the DARPA crew modulate the difficulty of the challenge. The terrain would be rough, but they would space out the vehicles, so each team would only have to deal with an unmoving world—avoiding rocks, cactus, barbed wire, and the like, not other vehicles. Maybe someday the cars could tackle more complicated tasks like negotiating traffic. For now, just going from one point to another at a reasonable speed was tough enough.

The American desert, though, presented unexpected problems. For all its crazy history, DARPA had never had to worry about running near the burial grounds of Native American tribes. Negron had to convince LA Water and Power that the robotic vehicles wouldn't topple their towers, and work with law enforcement to close the necessary roads and railroads in the area. And he had to negotiate with one of the Mojave's most established populations.

Long before Tony Tether had called this place his racetrack, before Chuck Yeager broke the sound barrier in the skies overhead, before the white men came through on their search for gold, even before the Mojave and Chemehuevi tribes settled there, the sands of the Mojave were home to the desert tortoise. By 2003, habitat loss, disease, and hungry ravens drawn into their territory by human activity had turned the ponderous creatures into an endangered species. That status, at least, earned the tortoise an ally in the US Fish and Wildlife Service, which had no intention of seeing them crushed by a passel of off-the-leash robots. To placate the bureaucrats, Negron had to draw up an environmental action plan that would keep the animals safe, going beyond "If one's on the road, have somebody move it." The problem with picking up a desert tortoise, he learned, is that if you scare or stress it, it tends to pee. In a part of the world that gets five inches of rainfall a year, that lost hydration can be a death sentence. Even touching the animal is a violation of the Endangered Species Act. So Negron pledged to bring in a team of wildlife biologists to survey the course before and during the race, ready to fence in and guard any tortoises who wandered near the route, and with the training and legal right to move one if absolutely necessary.

When he wasn't haggling in the desert, Negron was at DARPA HQ, building a new kind of command-and-control system. He wanted a chase car following each robot through the desert, so his crew would need a way to track every vehicle. They would need the ability to remotely stop the robotic cars in case they went rogue, or to keep them from hitting one another. And they wanted to be able to start them remotely as well, for situations where they were just pausing a vehicle momentarily. The result, Tether said, was as complex as any vehicle communication system that the US military was using at the time.

The most foreign challenge, though, was designing an off-road race course. That's how Negron ended up on the phone with Sal Fish. A Los Angeles native in his early sixties with swept-back white hair and a bushy mustache venturing just beyond the corners of his mouth, Fish was something of a desert racing legend. He'd spent several decades in charge of SCORE International, the body that runs some of the world's most popular and grueling off-road races. He knew everything about creating courses that would push vehicles to their limits, and nothing about the Defense Advanced Research Projects Agency. So when his secretary told him a man from DARPA was on the line, Fish thought it was a prank call from one of his friends. Then the guy started talking about his plan to send a fleet of vehicles—without drivers inside—careening through the desert. "I have no idea what you mean by 'autonomous vehicle,' and I don't even think I can spell it," Fish told him. But he was intrigued.

After a trip to Washington to meet with Tether, Fish became an enthusiastic member of the team, scribbling notes onto a yellow legal pad in meetings full of people using laptops. "I had no clue what the hell these vehicles were going to look like, what they were going to do," Fish said. But DARPA had asked him for a course worthy of a Grand Challenge, and he would oblige. He explored the possible paths between Barstow and Primm, looking for a route that would include open terrain, elevation changes, hairpin turns, and as many dangers as possible.

As the details came together, Negron decided it was time to give the public more details than "autonomous vehicle race." He organized an informational session for February 2003, inviting anyone interested to come learn more about the rules, ask questions, propose ideas, and link up with partners and sponsors. Negron rented out the Petersen Automotive Museum in Los Angeles, thinking it a good venue for creating the future of driving.

As he made his way through LA traffic to the Petersen that February morning, Tether was sweating the small stuff. So detail-oriented that one program manager wondered if he had an eidetic memory, he was not thrilled to find that Negron had taken over an entire museum—and planned for a huge event. He thought that maybe ten people would show up, that they could have made do with some pizza and beer. *What are we going to do with $2,000 worth of food?* Tether thought. He planned to give the leftovers to the local homeless population.

As he pulled up to the museum, a hulking concrete building at the western end of LA's Miracle Mile, Tether realized there wouldn't be much to donate after all. A line of people, four abreast, stretched down the block and around the corner. Half an hour before the doors opened, hundreds of people were waiting to hear more about the Grand Challenge.

Holy cow. We really might have something here, Tether thought.

The Geeks and the Govvies

STANDING IN THAT LINE OF HUNDREDS OF PEOPLE OUTSIDE THE Petersen Museum in Los Angeles were two young men fresh off a long drive from the San Francisco Bay Area. One was Randy Miller, a graduate student at Stanford studying construction engineering. The other was his friend from their college days at UC Berkeley.

A few weeks shy of his twenty-third birthday, Anthony Levandowski was already a veteran engineer and entrepreneur. He grew up in Brussels with his mother, a French citizen who worked for the European Union. At fourteen, he moved to Northern California to live with his father, an American businessman. He wanted to attend high school in America, he said later, because he thought it would better set him up for professional success. Even if he couldn't write in English when he arrived, he didn't take long to embrace American-style capitalism. At Tamalpais High School in tony Marin County, Levandowski sold candy to fellow students. After building a website for the school, he launched a business doing the same for companies around town. Before leaving high school, he had made enough money to buy a three-bedroom house (with some help from his dad and stepmother) near the Berkeley campus.

He kept up a web business called La Raison when he started at Berkeley, but realized the only way to keep up with larger competitors was to cut his prices or win over clients with hands-on customer service.

"There was no barrier to entry there," he told the university news site. "I don't want to be someone just providing a commodity at a low price." Levandowski preferred outthinking people to outworking them. His approach to problem solving was constant pursuit of the shortcut, the hack, the way to game the system and jump ahead. Like many in the nearby tech world of Silicon Valley, he saw this not as cheating, but as good engineering. If spending endless hours helping customers with mundane computer problems was the way to make his business work, he would find something else to do.

As an undergrad at Berkeley, Levandowski studied industrial engineering and taught himself whatever he wanted to know, which was mostly computers. He spent vacations reading manuals and chose his housemates based, in part, on their programming skills. He wasn't much for late nights in the library. "I've never done much homework," he told the college magazine. "I think it's pointless." But when he found work he deemed worthwhile, he was unstoppable. Sleeping, eating, and socializing became tertiary concerns at best. Robots tended to be one of those worthwhile pursuits. He'd always liked the moving machines, the way they took code from a computer screen into the world. "They made computers real," he said. And he was good with them. As a junior, Levandowski won a regional robotics competition with BillSortBot, a Monopoly money-sorting machine he made largely of Legos. He gave it purple antennae and big eyes, just for fun. He was glad to come in first, but especially glad to have beaten Stanford, a perennial front-runner in robotics work.

Levandowski went on to graduate studies in industrial engineering at Berkeley and a fresh side hustle, working with his friend Randy Miller to make portable electronic blueprint displays for use on construction sites. But his focus shifted when his mother called from Belgium to tell him about a competition she'd seen announced. She remembered her son playing with remote control cars when he was little, and figured he'd like the chance to build one that could drive itself—and win a million bucks doing it.

Soon after, sitting in a hot tub at Levandowski's father's house in Lake Tahoe, Levandowski and Miller brainstormed ways to tackle the race. They considered an autonomous off-road excavator. A motorcycle that drove itself could be cool. Undecided, they headed south to DARPA's meeting.

By 9 a.m. on the day of the Petersen meeting in February 2003, everyone waiting to hear more about this Grand Challenge had made their way to the second floor of the building. Judging by looks alone, this was nothing like the standard "industry day" DARPA hosted to kick off a new program. Tony Tether, Jose Negron, and the rest of the "govvies" wore suits. But while the representatives of the defense contractors came in the same buttoned-down uniforms, the geeks showed up in jeans, sneakers, and T-shirts. Waiting for things to get started, they milled about a large room filled with rare and classic cars, eating the pastries and drinking the coffee Tether had thought he'd be handing out on Wilshire Boulevard.

Negron took the podium first. He gave his introductory spiel (welcome, thanks, and so on) and laid out the goals of the day: to explain the rules and logistics of the Grand Challenge, answer questions, and help interested individuals form into teams. Negron, who had been fielding questions from eager potential racers for months, made it clear just how nebulous the whole affair still was. "I have got one group on the left that says, 'This will never be done,' and I have got another group that says, 'I have got a car built already,'" he told the crowd. "Somewhere in between there is the truth. And we are going to figure that out together."

Then came the rules, as brutal as they were simple. DARPA would issue each team an emergency stop system. This bit of hardware, to be installed on the vehicle, would let race officials kill the engine with a remote control, as a backup safety measure. The vehicles would leave the starting line one by one, in five-minute intervals, to keep them spread out. The Challenge was robot vs. desert, not robot vs. robot. (That put the kibosh on one competitor's desire to use a roof-mounted cannon.) Once a vehicle got going, its creators would not be able to help it in any way. If it ran off course, lost a sensor, got stuck, or anything, it would have to solve the problem on its own. Each would have just ten hours to complete the course, which could be as long as 250 miles. To get to the end, the vehicles would follow a series of GPS waypoints, about one hundred yards apart on average, leading them from Barstow, California, to Primm, Nevada. DARPA wouldn't reveal those coordinates until just two hours before the flag dropped on race day, set for March 13, 2004. That way, Tether calculated, the teams would have to prepare their vehicles for everything. And the showman in him liked the added touch of mystique.

Sal Fish hadn't finalized the race course yet, but he was far enough along to take the podium and show the competitors what sort of terrain awaited them. He had donned a suit for the occasion, but his unruly mustache, his left-out-in-the-sun-too-long look, and his colloquial, rambling manner of speech made clear he was no government guy. No one would have mistaken him for a geek, either. Fish was here to put the fear of the desert he knew so well into these novices. He clicked through a slide show catalogue of horrors. Dips. Gullies. Water deep enough to drown a Jeep's engine. Forty-five-degree inclines. About a fifth of the course would be on trails, but those could be made up of loose sand or run along narrow ridgetops. "This is a real tire-eater," Fish said about a photo of a trail studded with volcanic rocks. "Those are very sharp rocks, and it's very challenging. Next slide."

During a lengthy Q&A session, Anthony Levandowski went to the microphone multiple times, posing questions that revealed a penchant for creative thinking. Can you have two vehicles, exchanging location information? (No, the answer came back, but "good try.") During the two hours between the reveal of the course and the race start, can you use a plane or drone to aerially survey the area? ("We'll get back to you.") But leaving the Petersen, the young engineer was still unsure of his approach to the race.

Then, on the freeway heading back to Berkeley, a pack of motorcyclists streamed past his car. Watching them roar by, a few gears clicked in Levandowski's mind. Motorcycles were quick and agile. Compared to a vehicle that took up the whole road, a narrower, more maneuverable two-wheeler would have extra room for error on those tight desert trails. These were the sorts of advantages that could get him closer to that $1 million prize, and no uncertain amount of glory. So he made up his mind. His entry to the DARPA Grand Challenge would be an autonomous motorcycle. He just needed to figure out how to make such a thing.

Relatively few of the people who had jammed into the Petersen Automotive Museum ran in the academic circles that had produced the artificial intelligence and robotics research that made Tony Tether think the Grand Challenge was possible. The world's most sophisticated robots might be able to speed down a well-painted, smooth highway. Others could crawl over unmarked, rugged terrain. But none could take the speed off-road, which was why Tether wanted new people to come in and give the field a good kick forward. And those in the room who knew anything at all about robots surely took a blow to their confidence when they saw William "Red" Whittaker walk in.

Whittaker wasn't in Los Angeles to ask anyone's advice, find teammates, or woo an angel investor. He was there to learn what this Grand Challenge was all about, and to make sure that if anybody won the million dollars, it would be him and his beloved home, Pittsburgh's Carnegie Mellon University.

The son of a chemist mother and a father who sold explosives (both of them pilots), Whittaker grew up in central Pennsylvania, a few hours east of Pittsburgh. He was loud, brimming with confidence, and impatient to the point of rudeness with anyone who asked a question he deemed unworthy of his time, thought 7 a.m. was too early for a meeting, or used the word "maybe." He was prone to telling students perplexing things such as *This project is like a freight train. You've got to grab on, and it'll rip your arms off!* But his favorite phrase was "rock solid," which happened to be an apt physical description for a man who was six-two and built like a redwood. Whittaker left Princeton during the Vietnam War to join the Marines, where he boxed and played football.

After his service, he finished at Princeton, then moved to Carnegie Mellon University for a master's and PhD in civil engineering. But Whittaker didn't want to build bridges. He wanted to build robots, a passion he'd maintained since he put together creations as a kid, using parts he collected from a junkyard near his house. "I sought something that would dent the world, that I could do with my own hands, that would happen in my time," he said. He thought the emerging study of robotics would provide the space for him to land his hammer.

Then and now, Carnegie Mellon was maybe the best place in the world to build robots. Established in 1900 by Andrew Carnegie, who brought the steel industry to Pittsburgh, the technology-focused school started the country's first dedicated robotics institute in 1979. It was a prescient move at a time when researchers were just beginning to imagine how they could shift computerized intelligence from their labs into the real world. Whittaker joined the new institute as a freshly minted doctor of engineering.

That same year, a reactor at the Three Mile Island nuclear power plant near Harrisburg, Pennsylvania, partially melted down. America's precursor to Chernobyl presented Whittaker with an opportunity. Thirty years old and still in possession of the red hair that gave him his nickname, he landed a government contract to create robots that could venture into the devastated reactor and assess the damage. "I took the money and burned around the clock for about six months, and delivered the goods," Whittaker said. His creations spent four years wandering the building's dark, flooded basement, snapping photos, recording video, drilling into concrete walls, and measuring radiation.

These were among the first entries in a discipline Whittaker called "field robotics." Instead of sitting on factory floors, bolting together car bodies that came to them on assembly lines, these machines engaged with a dynamic, uncontrolled world. In 1986, Whittaker led the foundation of the Field Robotics Center, a new arm of CMU's Robotics Institute. His first attempt at what you might call an autonomous car was called Terregator. Equipped with cameras, sonar, and a laser range finder, the six-wheeled robot resembled a refrigerator. For some tests, it navigated sidewalks at walking speed, connected to its brain by a hundred yards of cable that ran out the window of the fourth-floor computer lab.

Whittaker could write software, but his genius lay in his innate sense of how to make something robust and reliable. "He gets how things work," said his longtime colleague David Wettergreen. Whittaker thought at a *systems level*, considering how the components of a machine would work together. He built one robot after another that didn't just survive battering conditions, but thrived in them. Perhaps incidentally, he tended to make them in his own image: hulking, with an implacable sense of mission. In the mid-eighties, when Carnegie Mellon started its pioneering Navigational Laboratory autonomous driving program, Whittaker did much of the work to turn that blue Chevy panel van into NavLab 1. The young Robotics Institute didn't have the space to do the work inside at the time, so Whittaker spent hours in an outdoor loading bay in the Pittsburgh winter, hooking up cameras, radars, laser scanners, and computers. Through the end of the millennium, he, with his colleagues and students, built dozens of remarkable robots that did the sorts of things feeble human beings have trouble with, in the sorts of places that tend to kill us. The machines climbed into volcanoes in Antarctica and Alaska. One found steady work inspecting between flights the seventeen thousand tiles coating the underside of the Space Shuttle. Others mapped abandoned coal mines and prepared to chart the solar system. When he wasn't on campus or in some far-flung place with his latest invention, Whittaker worked his cattle farm outside the

city, which he had bought in the 1990s to give him something physical
to do besides maintaining his physique lifting massive weights.

Whittaker was fully plugged into the robotics world and had done
plenty of work with DARPA, so when Tony Tether announced the
Grand Challenge, it wasn't long before he caught word. At first, the idea
of a race struck the roboticist as frivolous. His robots were tools, not
toys. But the former Marine liked the military angle, the chance to take
men and women out of harm's way. Plus, he had done much of the basic
research that DARPA now wanted to convert into a viable product. His
ego and his fierce competitive side hated the idea that someone else
might filch the baton for the sprint to the finish line. And, at a deeper
level, he saw a chance to prove to the whole world what robots like his
could do.

His peers weren't so enthusiastic, and the school's administration
didn't love the idea. It wasn't just that there was no clear research angle,
no paper that would come out of making a racing robot. At a university
with a relatively small endowment, funding was a precious thing. No
one knew how much it would cost to compete in the Grand Challenge,
but it sounded expensive. And they knew you only got the million dollars if you won. It was like playing a lottery where the jackpot might not
cover the price of a ticket.

"Some people think about stuff like that," Whittaker said. To him,
such worries sounded like excuses. As for the administration's reluctance, well, he had never let hand-wringing and paperwork stop him
from doing what he thought needed doing. Years earlier, a close friend
of Whittaker's had died climbing a mountain in South America. He flew
down, made the climb himself, and came back down carrying his friend.
One weekend morning in the 1980s, he showed up a few minutes late
to a meeting with a grad student—a rare occurrence explained by the
gruesome road rash on his arms, dripping blood onto the table. It was
only later that the student learned Whittaker had been biking through
the park that morning, going full speed, when he came around a corner
and found himself on course to collide with a woman pushing a baby

carriage. Whittaker juked out of the way and hit a low stone wall, flying over it and down into the woods.

He scoffed at the hesitation to join the Grand Challenge on the part of CMU's higher-ups. "Life isn't 'Mother, may I,'" he said. Plus, he'd always liked programs with clear mission statements, and it didn't get much clearer than *get there first.*

On March 13, 2003, Whittaker wrote a blog post declaring he was in, giving DARPA its first official Grand Challenge entrant. In the opening entry of the online journal he would use to track his quest, he made clear Tony Tether wasn't the only one with a flair for the dramatic. "Our run in the LA Vegas robot race will be history a year from tonight," he wrote. "The race defies prevailing technology, and many hold that the challenge prize is unwinnable in our time. We race to put winning technology in the winner's circle, to engage a generation of youth, to inspire ourselves and our world, and to transform the view of what is possible. . . . Until one is committed, there is hesitancy, the chance to draw back, always ineffectiveness. Concerning all acts of initiative and creation, there is one elementary truth the ignorance of which kills countless ideas and splendid plans: that the moment one definitely commits oneself, then providence moves too."

A week later, Whittaker laid out his game plan. The first month would be putting the basic pieces together: building a team, figuring out funding, and settling on a general technical approach. The team would spend the summer building all the subsystems, and the autumn putting them together into a vehicle. Whittaker knew the real challenge with any robotic system came once you had an integrated system, when you realized just how many bugs had worked their way in. Starting in November, the team would spend every day hunting down and exterminating those bugs. By March 2004, they'd be rock solid.

First, though, Whittaker needed that team. And the roboticist who was destined to help lead it was a little hard to find at the moment.

———

Nearly five thousand miles from Pittsburgh, Hyperion was inching across the desert. The ten-foot-tall bot ran on four bicycle wheels and carried a tray of solar panels that provided power for its cameras and laser scanner. It was funded by NASA, linked to the search for life and water in Chile's Atacama Desert. The most arid place on Earth was a good proxy for Mars.

Chris Urmson was watching his creation inch along at less than a mile per hour when Red Whittaker showed up. Where the older roboticist was a torrent of bluster and action, the twenty-seven-year-old Urmson was contemplative and soft-spoken, with a round face and sandy blond hair that slipped onto his forehead when he was excited or working hard. After racking up science fair awards as a high schooler in Canada, Urmson had won scholarships, studied computer engineering at the University of Manitoba, and graduated with a nearly perfect GPA. He was considering a career in bioengineering until he saw a poster featuring Dante, an eight-legged robot that had rappelled into an Antarctic volcano to study the roiling lake of lava within. The poster advertised the research being done at Carnegie Mellon University. Dante was a Red Whittaker project. "That looks cool," Urmson thought. He headed to America.

Five years later, Urmson had helped design robots to search Antarctica for meteorites and assemble space stations in orbit. When DARPA announced the Grand Challenge, he was living out of a bright yellow tent in a place whose native population consisted mostly of microbes. On the verge of finishing his PhD, he was trying to make this solar-powered robot crawl a single kilometer without requiring human help.

Whittaker was one of Urmson's advisors, and he dropped into the desert to check in on the project. First, he addressed the team: *Don't come back until you've covered that kilometer.* Then, he pulled Urmson aside, said that he was building a team for the DARPA Grand Challenge, and that he wanted Urmson to help him win. Joining up would mean delaying his dissertation and the start of his postdoctoral career, but the idea of a race across the desert hit the same nerve touched by that poster of Dante years earlier. It would be a tremendous challenge, a great way to advance the technology to which he was dedicating his life. And moving at 50 mph was the sort of thing that had brought the young engineer to Carnegie Mellon: It sounded cool. He was in. Urmson returned to Pittsburgh and to his wife, who was pregnant with their first child. He joined what Whittaker, never self-effacing, dubbed the "Red Team."

Because all armies need foot soldiers, Whittaker ran an undergraduate course called Fundamentals of Mobile Robot Development, whose students would work on the Challenge. Not that his grunts couldn't rise up. Whittaker believed anybody could make himself useful, and was willing to give even the greenest freshman a chance. Those who proved themselves tended to end up in leadership positions.

His two dozen or so recruits spent the summer doing preliminary work with a red Jeep Wrangler better known as NavLab 11, the latest (and ultimately final) descendant from the blue Chevy panel van. The team collected maps of the Mojave and tested different kinds of cameras, radars, and laser range finders for their ability to spot obstacles at different speeds and distances.

To bone up on the basics of racing strategy, Whittaker connected with Chip Ganassi, a Pittsburgh native and motorsports bigwig whose teams had competed in many of the world's elite endurance races. Over the summer, Ganassi attended a few team meetings. He explained how his teams ramped up testing, how they practiced in conditions that approximated as much as possible the race itself, and the importance of getting every little basic detail right. He taught them an old racing maxim that matched Whittaker's emphasis on reliability: *To finish first, you must first finish.* Ganassi didn't know much about software, but he knew that zeros and ones didn't matter if you got a flat tire or your engine overheated.

At this early point in the project, the rather gung-ho students had considered making their own vehicle. Hearing them debate making a giant ball that would roll across the desert, and a tricycle with wheels seven feet in diameter, Ganassi realized these kids didn't appreciate what off-roading entailed. At one meeting, he asked the group how many of them had raced in the desert. No surprise, nobody had. He asked how many had driven in the desert. A single hand went up. He tried again: "How many of you have even *been* to the desert?" Another hand, maybe two. That, at least, Ganassi could fix.

In late July, Whittaker and a few of his students found themselves sitting in Ganassi's private jet, flying from Pittsburgh to Southern California. Out in the Mojave, Ganassi introduced them to Rod Millen, a champion rally, hill climb, and off-road racer. Millen walked them around a shop where he was developing beefed up vehicles for the Marine Corps, and took them for a spin. Bouncing up and down and knocking into each other, even at a relatively sedate 20 mph, the desert delegation grasped what they were up against. It wasn't just making a machine that could see and sidestep boulders. It was making a robot that could race—not just drive—across this terrain without damaging any of its equipment, much of which had been designed to work in the comfy confines of a temperature-controlled server room.

All the while, Whittaker dialed for dollars, cajoling everyone he knew from a lifetime in robotics, and plenty of people he didn't, to come on as a sponsor, donate money, or provide expensive parts. Intel provided computers. Caterpillar donated laser scanners. Google sent cash, Goodyear tossed in run-flat tires. Boeing offered not just access to its machine shop, but the full-time help of two of its engineers. Ultimately, Whittaker pulled in about $500,000 in cash and millions more in donated tech and man hours, with support coming from more than forty companies. No other Grand Challenge team would marshal anything near that level of support. Still, it was less than what Whittaker had been hoping for, and just barely enough for what he had in mind.

In September, with a basic game plan in place, it was time to start building the robot. After that rough ride through the desert and a fair bit of haranguing from Ganassi, the Red Team knew they needed something tough and uncomplicated. Whittaker bought a 1986 Humvee from a local farmer for about $20,000. The team called it Sandstorm, because it sounded cool. And they started turning it into one of the most sophisticated driving machines the world had ever seen.

Building an autonomous vehicle is a matter of replacing, with machinery, the three things humans need to drive: the hands and feet that operate the controls; the senses that perceive their surroundings; and the brain that turns data into decisions.

The first bit is mostly mechanical. The Red Team would need actuators, simple mechanical tools that convert the electronic impulses of a computer into physical movements. (The Humvee's age proved helpful here: Hailing from a mechanical era, it had virtually no electronics to get in the way, not even antilock brakes that might try to interfere with an invading computer's demands.) For these and other mechanical necessities, they went grave robbing, pulling apart old robots and grabbing whatever they needed from within. Carnegie Mellon's Robotics Institute had a long-standing unspoken policy of recycling old projects into new ones, and the Red Team's limited budget outranked someone else's claim on posterity. Whittaker's army fitted Sandstorm with actuators to manipulate the steering column, work the brakes, and pull the throttle cable. In six weeks, the team had Sandstorm up on blocks in the shop, testing their computer's ability to control the vehicle.

Replacing a human driver's senses is more complicated. This was Chris Urmson's specialty, the thing he was doing with his sedate robot in the Chilean desert. Urmson's work and decades of Carnegie Mellon research had found that a quartet of tools could let a robot move through the world. The first was GPS, which Sandstorm could use to follow DARPA's waypoints from the starting line to the checkered flag. The Red Team chose a unit that doubled as an inertial sensor, so Sandstorm would know when it was speeding up, slowing down, or turning, much like a person's inner ear.

The three other tools were Sandstorm's best hope for navigating the one hundred yards of hazard-laden terrain that lay between one way-point and the next. Each tool brought its own mix of advantages and shortcomings. Radar was robust, cheap, and could detect obstacles hundreds of yards away. It wasn't much for detail, but could see through dust and fog, so the team stuck a device near the front of the Humvee. Stereo cameras could see everything just fine, and were rather helpful as long as they came with sophisticated software that could translate a two-dimensional photograph into something a computer could understand. What a human does without thinking—say, distinguishing a rock from its shadow—could easily beguile a machine. Glare and changing lighting conditions—unavoidable when working outdoors—made things even tougher. But the cameras could provide valuable information, so Sandstorm got a pair of them, each the size of a can of tennis balls.

Then there's Lidar. Lidar works much like radar, but instead of emitting radio waves and listening for their return after they hit surrounding objects, it fires out pulses of infrared light and measures how long they take to bounce back, using the feedback to create a virtual map of physical space. (The word was originally a portmanteau of "light" and "radar," but now is considered an acronym for "light detection and ranging.") The system could provide an obstacle's location down to the millimeter. It wasn't much bothered by the movement of the sun, and conveyed data in a format easily understood by a computer. So Sandstorm got three short-range Lidars and one that could see seventy-five yards out. Sitting on a mechanical arm, the long-range unit looked like a cocktail shaker and produced sixty thousand data points a second.

To keep the cameras and the main Lidar pointed the right way on a course sure to be full of climbs, drops, and tight turns, Whittaker had his students build a gimbal, a device that would swing the sensors this way and that to piece together a grander view. It would measure Sandstorm's movement and adjust the sensors' positions accordingly. If the Humvee was climbing a hill, the laser could swing down to keep pointing at the road instead of uselessly shooting into the sky. The team housed the gimbal and Lidar in a beach ball–sized eyeball with a flat glass front that became Sandstorm's distinguishing feature. To keep dust from becoming a problem, the students stuck a small keg on the Humvee and filled it with water. They ran hoses from the keg to each vulnerable sensor, so Sandstorm could spray its eye clean.

The next challenge was mounting all these sensors and connecting them to the onboard computers. One enthusiastic undergrad figured the best way to get everything up high, for the best vantage point, was to saw off the roll bars that supported the soft roof, then build up from a flat bed.

"We came in in the morning, and the top was missing," said Kevin Peterson. A senior, Peterson had signed up after seeing Whittaker's poster for the course (featuring a laser-shooting dune buggy flying through the desert) and became a key team member. Whittaker had always empha- sized personal responsibility, giving general goals and expecting people to take action and get results. The upside was that people like Peterson could emerge as natural leaders. The downside was that every so often, somebody decapitated a robot. "It wasn't exactly what we had in mind," Urmson said. Committed to a convertible, the crew ripped out the seats and dropped an 800-pound aluminum box the size of a walk-in freezer in their place.

Inside that box went the most powerful computers the team could get their hands on and make work in their off-roading robot. They formed the brain, running the software that determined how the Humvee would act. They churned through the Lidar, camera, GPS, and radar data that streamed in through thick black cables, some identified with printed labels, most with scribbles in black marker. They compiled the info into a coherent image of Sandstorm's surroundings, divided into two general groups: flat and not flat. Then they combined that intel with preprogrammed maps and calculated the path most likely to keep them on safe ground and moving toward the finish line. Finally, they fired the resulting orders to the actuators working the steering wheel, throttle, and brakes.

Over the course of the fall of 2003, the members of the Red Team wound up in one of two groups. In the first were the casual participants, the kids taking Whittaker's class and doing their share, but also tending to other commitments in their academic and social lives. In the second were the dozen or so diehards, for whom the Grand Challenge became all-consuming. Like Peterson, who would wake up at four or five every morning, head straight to work, and go until he couldn't. Or Yu Kato, who took charge of building the gimbal. He brought his rice cooker to campus and started spending nights in the Planetary Robotics Building, working and reworking the complex interplay of actuators, position sensors, and fiberoptic gyroscopes.

Phil Koon, one of the engineers Boeing sent over, suffered a bit of culture shock when he relocated with his family from Huntsville, Alabama. He arrived in Pittsburgh just as Sandstorm's conversion began, and soon realized students of the Red Team moved fast and without fear of getting something wrong. At Boeing, you didn't just decide to cut a vehicle in half. But Koon, who had unsuccessfully pushed his employer to field its own team, dived in. Before long, his wife handed him a sleeping bag and said that if he was going to come home after midnight and wake up their young children, he had better not come home at all.

One night after another, Spencer Spiker downgraded his hopes for getting home in time for dinner to getting home before sunrise. A former Army Ranger with mechanical experience and time to kill, Spiker had joined the team as a volunteer. His wife soon found herself, along with Koon's wife, in the ever-growing club of "Red Widows." Membership involved watching your spouse disappear into Whittaker's lab like a soldier shipping out to war.

Whittaker encouraged this kind of dedication. "There is no Easter, no Fourth of July, no Sundays," he said. Taking Saturdays off, apparently, didn't occur to him. And he didn't mind stealing his students' attention away from their other studies. "I don't want to overdo it," he said with typical bravado, "but if I ever caught them going to class I'd toss them off the team."

———

Three months after buying the stock 1986 Humvee, the Red Team was ready to leave the confines of the Planetary Robotics Building. Whittaker took them to an abandoned steel mill site along Pittsburgh's Monongahela River. As the steel industry cratered, the mill had shut down in 1997, its smokestacks demolished. That gave the team 178 acres of pot-holed, rubble-strewn roads on which to test, roads that also tended to flatten the tires and wreck the suspensions of their personal cars. It was a handy approximation of what their robot would face in the Mojave, and remote enough that if Sandstorm went wild, it was unlikely to steamroll an innocent bystander. They set up shop in the old railroad roundhouse, a big building with roll-up doors once used to repair the locomotives that hauled Pittsburgh steel. Whittaker called it a playground.

It was also pretty much what you'd expect of a place that had started making steel a century before Congress established the Environmental Protection Agency. Low shrub grass was the only thing growing. When it rained, a sheen of oil topped the puddles. If the roboticists ever got clear of the diesel Sandstorm guzzled and belched, they instead inhaled the stink of a toxin-filled river. There were no nearby bathrooms, and weird rashes were common. "It was super-toxic, but a beautiful, beautiful test site," said Matthew Johnson-Roberson, another undergrad-turned-zealot.

Years later, these members of the Red Team wouldn't be able to recall the first time Sandstorm drove itself, and for good reason. Carnegie Mellon had been making autonomous robots for more than two decades by this point. There was no question of whether they could get a Humvee to roll around on its own. The question was, could they make it roll for hundreds of miles, at speed, around obstacles, without a single failure come race day? Whittaker had been around long enough to know just how much could go wrong when you took a robot out of the lab and into the world. He wanted his students to know it, too. So he gave them a new assignment, one that would determine their grades for the semester: Prove their robot was rock solid by making it drive 150 miles, in one shot.

What the team would remember was standing around the steel mill site in December of 2003, watching Sandstorm endlessly circle the oval track they had marked off on the flat, concrete foundation of a demolished building. Listening to Chris Urmson counting off laps over the radio. Eating goulash from the nearby Hungarian hole-in-the-wall restaurant. Huddling around the space heater they installed in the "treehouse" they'd built to give them an elevated view and some cover from the snow.

And watching their robot fail again and again and again, each time in a seemingly novel way. The GPS would cut out. The steering or throttle controller would reset itself. A slight bump would upset a hard drive. Sometimes, Sandstorm just stopped. Other times, it would veer off the track, forcing someone to hit the emergency stop before it crashed through a fence. Or it crashed through the fence anyway. Each time, the young roboticists would look for what had gone wrong, fix it as best they could, put Sandstorm back on the track, and grimace as Urmson restarted his count.

— 3 —

Orphans Preferred

IN BERKELEY, ANTHONY LEVANDOWSKI WASN'T THINKING about how to make a car navigate without a human inside. He was trying to make a motorcycle stand up without a human in the saddle. As he'd realized during his drive north from the meeting at the Petersen Museum, motorcycles were quicker and more maneuverable than something like a Humvee. The idea was sure to stand out, which could attract the attention of reporters and potential sponsors. Plus, the notion of a motorcycle that drove itself was just plain cool.

That appeal helped Levandowski's recruiting efforts. He pulled in friends and friends of friends, most of them from UC Berkeley, pitching the motorcycle partly as a way to win the Challenge and partly as a way to stand out. If someone else won, his team wouldn't be one more group with a laser-clad dune buggy. They'd be the guys (and they were all guys) who had turned a motorcycle into a robot. The very thought was exciting for Dezhen Song, who was working on his PhD when Levandowski brought him aboard. The two had worked together under Ken Goldberg in Berkeley's Industrial Engineering and Operations Research Department. Goldberg, one of the country's premier roboticists, would soon realize that half his PhD students had joined what Levandowski dubbed the Blue Team, the term that designates friendly forces in war games. They divvied up the work, figuring out who'd write software, who'd do the welding, who'd make all the sensors work together. Levandowski asked his friend Ognen Stojanovski, who was getting ready to abandon engineering in favor of law school, to negotiate with UC Berkeley, hoping to get support in the form of development resources and lab space, while looking to retain control of any resulting intellectual property. It was an early sign of Levandowski's nose for opportunity, his habit of keeping an eye toward the future. Transportation accounted for an enormous amount of the economy, and you didn't have to take over much of it to fill your bank account. He wasn't sure what might come of making a motorcycle that didn't need a human rider. But he was looking far beyond Tony Tether's million dollars.

Levandowski put himself in charge of his team. He provided the pizza and burritos; he paid for the parts and the computers and the gasoline they poured into the motorcycle with a big yellow funnel. His team worked out of his house in Berkeley. They typed code under the vaulted wood ceiling of the living room and welded their machine together in the garage that connected to Levandowski's bedroom. In January 2004, he registered with the State of California, under his name, his latest company: Robotic Infantry, Inc. Of course, ownership of the team's work didn't mean much if they couldn't make anything.

Equipping a car to translate computer commands into physical actions means hooking up motors that do what a human driver does: press the throttle, apply the brakes, turn the steering wheel. A motorcycle rider, though, does more than control inputs to the machine. She shifts her weight left and right, doing much of the work to swing through turns. Levandowski's crew wouldn't have the benefit of an intelligent rider with movable mass. One possibility was a gyroscope, which provides balance via a wheel spinning on an axle. In the 1960s, auto industry designers had used the device to keep two-wheeled concept cars upright. Yacht manufacturers deployed gyroscopes to keep their vessels steady in choppy waters. The Blue Team built one in Levandowski's kitchen, but decided it was too bulky for their small bike. Besides, Levandowski didn't want a complex mechanical setup that couldn't work for lots of vehicles. He wanted a software solution. An answer written in code was more elegant, and easier to scale into a commercial system.

Levandowski understood the basic principles of staying upright on two wheels; he had ridden dirt bikes when he was younger and used a scooter to get around Berkeley. But to make a motorcycle do the same on its own, Levandowski and his team had to apply exactly the right amount of power and turn the wheel exactly the right number of degrees, at exactly the right time. Each of those "exactlys" changed depending on the speed of the bike, the terrain on which it was rolling, its position, and its trajectory at any given moment. The young engineers had to figure out all of this despite sensors that threw up false readings or disagreed with one another, and being hobbled by how much computing power they could strap onto their vehicle. Weight was such a concern that they measured their computing power not just in processing capacity, but in pounds.

The team started with an old Honda motorcycle, then switched to a 125cc Yamaha, then to a 90cc Yamaha with an automatic transmission, saving them the work of tussling with a clutch and gear selector. They called their creation Dexterit, and later went with the more evocative Ghostrider. But from early on, Levandowski wanted to test new approaches without taking the time and energy to build them into the full-sized bike. Which is how the Blue Team ended up in his driveway, passing around the remote control that governed the ten-inch-long model motorcycle Levandowski had bought online. They used it to see how changes like moving the center of gravity around changed its dynamics as it rode between the tufts of grass growing through the cracks in the pavement, leaning this way and that. They studied how it stayed upright and what made it fall over. They went bigger, too, putting an inertial measurement unit and a few accelerometers onto a bicycle, to create a mathematical understanding of how humans balance. Somebody would pedal around the block while the others watched the data returns on Levandowski's laptop. What sorts of physical forces went into a turn? Where did the wheels point, how fast did they spin?

When it was time to see what their zeroes and ones could do, Levandowski covered the motorcycle with a cornucopia of gizmos: inclinometers, accelerometers, a GPS, motors to work the throttle, brakes, and steering. He stuck a piece of plywood just behind where the seat had been, as a stand for the laptop that controlled everything. Much of it had come for free, or at a steep discount. Without official support or access to Red Whittaker's Rolodex, Levandowski had cold-called hundreds of potential sponsors. His first steps made clear he had minimal marketing savvy. In May of 2003, he sent this letter to BMW:

> *Hello,*
>
> *I am a UC Berkeley Graduate student in the Electrical Engineering department and I am trying to reach BMW motorcycles corporate marketing.*
>
> *We are building the world's first autonomous motorcycle and would like to talk about a possible collaboration with BMW.*
>
> *Best regards,*
>
> *Anthony Levandowski*

He never heard back, of course. But with everyone he got on the phone or met in person, Levandowski flashed a salesman's talent for winning skeptics to his worldview. He landed help from defense contractor Raytheon, Hewlett-Packard spinoff Agilent, and sensor manufacturer Crossbow, mostly in the form of free equipment. Each deal lent the project credibility and connections, making the next one that much easier to lock down. And he honed his pitching strategy. In an October letter to chipmaker Nvidia, Levandowski talked up what his team had accomplished so far, which milestones they hoped to hit, other sponsors who had come aboard, and how much media coverage the team was receiving. Indeed, for any reporter covering the Grand Challenge, including Levandowski and his computer-clad motorcycle was an obvious choice. He was featured in *WIRED* and the *New York Times* as a bootstrapped, kookier foil to Red Whittaker's near-professional effort.

The writers harped on the nuttiness of the project, but also credited the twenty-three-year-old for trying something so difficult and yes, cool.

Levandowski ultimately scraped together about $30,000 from sponsors and a pile more from his own savings. He understood that success in the long run called for sacrifice now. He was fine driving his hand-me-down 1991 Nissan pickup truck, just as he was fine sleeping in the living room of his own house so he could rent out the master bedroom.

As the team put all these tools together, they grappled with the intricacies of robotics. The team members, including Levandowski, had worked on various robotics projects, but the complexity of the autonomous motorcycle introduced the sort of stuff that made Red Whittaker's decades of experience crucial for the team building Sandstorm. So Levandowski studied up. He went to trade shows, asking vendors why their products weren't working for him. More often than not, they'd explain he was using a tool the wrong way, or the wrong tool altogether. Over the course of 2003, he and his teammates started to master concepts and tools such as encoders, couplings, backlash, and compliance.

Even with all these lessons, as the year drew toward an end—while Carnegie Mellon's team was prepping for its 150-mile run—the motorcycle was still going just a few yards before toppling over. After nearly a year of work, the project had become all consuming for Levandowski. He was working sixteen-hour days, seven days a week. No one knew when he slept, if at all. With every fall of the bike, his frustration grew. His identity as an engineer hinged on his ability to solve problems, to make things work. But each iteration of the software, each reconfiguration of the sensors, produced the same result. The motorcycle would wobble and crash, wobble and crash. Not that Levandowski considered giving up. Abandoning it now would mean admitting he'd been a fool to try something so outrageous.

More surprising was that his small team proved loyal. They weren't getting paid, unless you count burritos and pizza. They weren't getting school credit. The project devoured time they could have spent studying. A laughable attempt to make a robot motorcycle didn't seem like a great way to burnish a résumé. But they stayed, in part, because of what Levandowski was doing. He believed in what he was selling. For all his gangly geekiness, magnified by his six-foot-six frame and manic gesticulations, his optimism was charming. He was fun to be around, free of doubts and full of bold ideas. His enthusiasm had an infectious quality. "I remember thinking, 'I hope he likes me and lets me work on this,'" teammate Charlie Smart said. Ognen Stojanovski said Levandowski tapped into the "think of what you'll tell your grandkids" vein. Of course this was hard! If it wasn't hard, someone would have done it already. That's what made it worth the effort. Each difficulty would just sweeten the ultimate victory and make the story that much richer.

Convincing everyone to stick around got a lot easier at the end of 2003, when the team uploaded yet another iteration of its software to the bike. Years later, Levandowski wouldn't be able to say what exactly had changed, but it was clear that this latest remix of how the motorcycle swung its front wheel and twisted its throttle worked. Not deftly. Not for long distances. But when the bike started to tip over, it would recover and carry on. Ghostrider was doing it.

With just a few months left before race day, Levandowski and his teammates rushed to take the bike from the equivalent of a five-year-old newly free of training wheels to that of a motocross champion. Levandowski would load the motorcycle into the back of his pickup and drive over to the Richmond Field Station, a UC Berkeley satellite campus twenty minutes away. The area, covering 170 acres between I-580 and the San Francisco Bay, housed the university programs that needed a bit of elbow room, including its Earthquake Engineering Center, Ergonomics Program, and Pavement Research Center. It was also home to Berkeley's Institute of Transportation Studies, which had been studying autonomous vehicles at the site since the late eighties. Where Carnegie Mellon worked on robots that coped with the human world, Berkeley's researchers focused on infrastructure-dependent solutions, like magnets embedded in freeway concrete and guiding the cars above. Their work, which had never moved past the demonstration phase, didn't fit Levandowski's approach. It required too much cooperation from outside parties, too much investment, too much time.

If not a helpful history, the Richmond Field Station had what the Blue Team needed: open expanses of grass, dirt, and pavement where Ghostrider could run free. Here, the motorcycle steadily improved—though it did smack into a sign warning passersby that this was an active autonomous vehicle test area. Levandowski and his teammates would load up various sets of GPS coordinates, sending the bike in circles, straight lines, figure eights, moving from one sort of terrain to the next, studying where it stayed up and where it stumbled, warning curious people in the area to keep their distance, watching spooked birds flee the scene, jogging behind the bike, putting it back on its wheels when it fell. Eventually, Ghostrider could even stand up straight without going anywhere, just swinging its front wheel left and right.

In early 2004, with race day set for mid-March, Levandowski decided it was time to take Ghostrider to the desert. The Blue Team put the motorcycle in the bed of his old Nissan and drove six hours south to El Mirage, a lonely town in the Mojave. They set the bike loose in the desert, and found that—with some tweaking—the software they'd programmed to work on grass and pavement held up on dirt and sand. During one especially impressive test run, Stojanovski was driving the pickup as a chase vehicle and, in his struggle to keep up through the bushes and cacti, scratched the side of the truck. But he got out feeling more buoyant than sheepish. Levandowski didn't care, either. Their motorcycle looked good.

Success in the Grand Challenge, of course, required more than a vehicle that could stay upright. An hours-long drive through the desert required that the motorcycle have the ability to find and stick to the road, to detect and avoid obstacles in its path. Levandowski had put this off to focus on the more pressing balance problem. Now he strapped on a camera that would let the bike make sense of the terrain around it. Side by side where the handlebars had been, they resembled the eyes on an overgrown bug. He and his teammates didn't have the time to program them, however, and they were mostly a nod to the idea of perception. Ghostrider could ride, and it could follow GPS directions. That ought to be plenty to impress the world.

As the March 13 race date approached, Carnegie Mellon's Sandstorm was coming together. The Red Team never quite hit that 150-mile run (and Whittaker didn't fail his students), but the effort had let them root out many problems with their system. Now their robot was cruising around the abandoned steel mill at speeds north of 30 mph. "Sandstorm performed rock-solidly," Whittaker wrote in his online journal in January 2004. "The machine is a beast. There were many 4-wheel slides through icy corners, and spinning wheels looked like snowblowers." But as its name indicated, Sandstorm didn't belong in the snow of a Pittsburgh winter.

At the end of January, the team loaded the Humvee, a pile of tools, spare parts, and computers into a tractor trailer. Whittaker stuck a stuffed groundhog in Sandstorm's driver's seat ("Red loves groundhogs," said Kevin Peterson) and sent it on a three-day, cross-country journey. A few team members, including Chris Urmson, flew to San Francisco, rented a pair of RVs and a couple of SUVs, drove over the Sierra Nevada, and rendezvoused with their robot at the Nevada Automotive Test Center. An hour southeast of Reno, the facility consisted of a few low buildings, including a garage, surrounded by about a million square miles of territory governed by the Bureau of Land Management. This was where the U.S. military and manufacturers of construction and farm equipment came to brutalize new tanks and tractors on the endless miles of rugged roads. If the old steel mill was a playground, this place was an amusement park.

Like so many zealots before them, the core members of the Red Team used the desert to purge everything that kept them from their holy mission. No family. No friends. No classes. Just the Grand Challenge. Sleep came not at night, but whenever they could snag a few hours, or couldn't go another second. "You just worked until you passed out," said Matthew Johnson-Roberson. As the lowest man on the totem pole, that sometimes left him sleeping on the tiny table in one of the RVs. Mostly, everyone took shifts in the beds above the front seats. Urmson laid out his sleeping bag in the garage.

The first night, they spilled diesel in one of their RVs, which didn't seem so bad by comparison when they later flooded one with the contents of the sewage tank. After that, they stopped using the RVs' water systems, including the showers. Every few days, someone would make the hour-long drive to the nearest grocery store and come back with a trunk full of trail mix, Saltine crackers, peanut butter, and bratwurst. There was a lot of bratwurst. The rare nights they finished (or at least paused) before the nearest restaurants closed, they'd pile into the SUVs and make the long drive, in exhausted silence, to Applebee's or Chili's.

Every day, they programmed their robot with a set of GPS waypoints that would tell it where to go. As it drove itself about, Chris Urmson or Kevin Peterson, who led the software work, would follow Sandstorm in a chase vehicle, looking for mistakes and rewriting some 750,000 lines of code as they went. They put cones and cardboard boxes in its way to check its obstacle detection skills. They set off smoke bombs to gauge how it would handle dust. When the sun set, it was back into the garage. Urmson and Peterson would keep working on their software. Yu Kato would work on the gimbal. Spencer Spiker and undergrad Nick Miller would attempt to fix whatever damage their teammates had done to the Humvee.

The crew started with 1.6-mile loops, improving the robot's ability to control itself, to stop clipping corners or drifting out of its lane. Soon, Sandstorm was hitting 40 mph. Then they sent it on longer drives about the desert, dealing with the sun's effect on the sensors (bright direct light is no friend to Lidar) and the engine's unhelpful tendency to die on steep slopes. They added a turbocharger to get more out of the old engine, and fitted it with a seventy-five-gallon fuel tank so the needle wouldn't approach empty.

Some of the routes followed the old trails of the Pony Express. Back in Pittsburgh, Whittaker dug up an ad for the famed mail service that read: "Wanted: Young, skinny, wiry fellows. Not over 18. Must be expert riders. Willing to risk death daily. Orphans preferred." He approved. "This is good language for our next recruiting poster," he wrote in his race log. The fifty-six-year-old roboticist had stayed home, still cajoling sponsors for much-needed cash and supplies, and preparing the Red Team's secret weapon.

The team had spotted a key problem early on. To complete the course in the allotted time and stay ahead of the field, Sandstorm would have to average about 14 mph. Its sensors, though, couldn't reliably see too far ahead. If the road suddenly turned, the speeding robot might not have enough time to detect the change and adjust its path. The way to keep the throttle down, they realized, was to tell Sandstorm where to go ahead of time. A map that showed where the road was wide and flat, or steep and narrow, or winding, or nonexistent—would let the robot stay on the gas as much as possible, while its sensors looked for small, unmappable threats like rocks.

The Red Team's mapping division started by downloading data from the United States Geological Survey for the region between the starting line in Barstow, California, and the finish in Primm, Nevada, along with data from a helpful local satellite company. Translating static maps into something Sandstorm's computers could read meant the team members had to comb through them by hand, labeling railroad tracks and rivers, noting elevation changes, recording road widths and turns. Then they translated them again, into guidelines for their robot. Go fast here. Avoid this spot. Slow down before this turn. But even with Whittaker putting on his typical pressure, they realized they couldn't cover the entire fifty-thousand-square-mile race region before race day.

Chris Urmson found another way. Along with the list of waypoints that would make up the Grand Challenge route, DARPA would designate how far a vehicle could stray from the direct line between any two points. Just as in baseball, where a base runner going from first to

second can't dash into the outfield to avoid being tagged out, the robots couldn't leave the designated corridor. The teams would get the
coordinates and the boundaries just two hours before the race started,
Tony Tether's way to keep anyone from mastering the course ahead
of time. But Urmson figured two hours might be just enough time to
comb through the data covering the route and give Sandstorm the fastest way to the finish line. Instead of a perfect map of the whole region,
they would produce a perfect path through it.

To do it, the mapping team wrote custom software that would pick
out the fastest route, accounting for things like road width and elevation
changes. Led by Alex Gutierrez, a grad student who served as a sort of
chief of staff for Whittaker (and TA for his classes), the team would then
divvy up the route. Each person would go through his section, verify-
ing the computer's decisions, clicking and dragging to adjust the line
that would determine the vehicle's trajectory. In the months leading up
to the Grand Challenge, each would take his place at a row of computers and practice the two-hour process over and over. Some days, they
would send their results to their teammates in Nevada, who'd plug them
into Sandstorm and see how they were doing.

By the start of March, with less than two weeks left before race day,
Sandstorm was coasting through GPS-blocking underpasses, sliding
around cattle guards, and running smoothly for hours on end. Months
of brutal work were finally showing results. A few days before the qualifying round that would lead them to the Grand Challenge, they would
finalize their code, run some last checks, and get a bit of rest before the
big day. But first, they wanted to do one more 150-mile run to shake out
any remaining bugs and see if they couldn't get the robot to run just a
bit faster.

The afternoon of March 4, 2004, the team gathered around the two-
mile oval track next to the garage where they did this sort of endurance testing. The first few laps, Sandstorm looked great, clocking nearly
50 mph on the straights and about half that on the turns. Eager to get
the test over with, Urmson told Kevin Peterson, sitting in one of the

SUVs and controlling the robot from a laptop, to up the speed. A few keystrokes later, the signal ran through the robot's bank of computers and down to the motor that did the job of the human's right foot. Sandstorm picked up the pace. Then, as it came around one corner, it juked right, then a bit left, then right again. Nick Miller saw the stutter step and called it out over the radio. Urmson noticed it, too, but decided to let the robot go another lap. If it did it again, they'd stop the test. Sandstorm circled the track, and when it got to that turn, it juked again. Harder this time. Attempting to correct, it veered too far in the other direction, pointing itself right at Miller. "Too fast! Too fast!" Urmson yelled into his radio. Miller ran for a nearby mound of dirt. Before he could throw himself behind it, he realized he was safe, because Sandstorm had changed direction again. Not left. Not right. Up.

Even in the Mojave, wet weather had proved a problem, as snowmelt came down from the mountains. A few days earlier, Sandstorm had gotten itself stuck in mud up to its belly, requiring a tow to freedom. Now, as it zigged and zagged, one of its wheels sank into the soft ground. The Humvee kicked into the air, and the high-riding weight of the electronics box pulled it end over end. Sandstorm landed on its roof. As the crunch hit their eardrums and the robot settled into the dirt, Peterson and Matthew Johnson-Roberson, sitting in the SUV, turned to each other with the same thought. *Whoa.*

Later, Urmson would realize that in programming the route, they had left two overlapping waypoints, which seemed to explain the juking. Not that it mattered. The qualifying round that would let them into the Grand Challenge was four days away, and Sandstorm was a wreck. The crash had annihilated the all-important gimbal and the sensors it wielded. Circling the fallen contender, Urmson did a quick evaluation of the robot's various parts. He used one phrase over and over: "That's fucked."

Boeing engineer Phil Koon was Whittaker's designated field team leader, by virtue of experience and seniority. That left him the job of picking up the phone. "Red," he said, "we rolled the vehicle." Whittaker came back with the calm of the Marine he never stopped being. He asked what was damaged, what the plan was to fix it, and how to execute it. That night, he wrote in his online journal: "The gimbal, dome and primary lidar are obliterated. The fin is heavily damaged, but may be salvageable. The power shelf broke loose. Shock isolation brackets are damaged. We don't yet know enough about Sandstorm." That was the windup. Then came the pitch. "We know the team. We'll pull this out."

After a few minutes of gaping at the carnage, the team got to work. They called in a wrecker to flip Sandstorm upright. They cleaned up the engine and got the oil out of the cylinders, then straightened the suspension, realigned the wheels, and fixed the body with a sledgehammer. By a stroke of luck, most of the damage, including a three-inch crack in the electronics box, was mechanical. If the electronics systems or the wiring that connected the sensors to the computers to the actuators had been damaged, the Red Team might not have had the time to recover.

The team's spare gimbal and sensors were sent by FedEx to California ahead of the qualifying round. Phil Koon called a Boeing colleague who worked near the West Coast shipping facility and asked him to drive the package to the desert. He did—on his motorcycle—then spent the night helping the team rebuild Sandstorm. In just two days, Sandstorm was running again. But without the time to do any serious testing, no one knew whether it was really back. The changes necessitated by the rollover called into question the validity of all their previous work. "In a moment," Whittaker wrote, "it went from a machine with hundreds of autonomous test miles to a newborn."

After four frantic days, the team ran out of time. They packed Sandstorm back into its trailer and sent it on the long drive south to Fontana, California, where they'd fight for their place in DARPA's Grand Challenge.

— 4 —

Convulsions and Smoke

WHEN THE INQUISITION REQUIRED HIM TO DROP HIS STUDY OF what the Roman Catholic Church insisted was not a heliocentric solar system, Galileo Galilei turned his energy to the less controversial question of how to stick a telescope onto a helmet. The king of Spain had offered a hefty reward to anyone who could solve the stubborn mystery of how to determine a ship's longitude while at sea: 6,000 ducats up front, and another 2,000 per year, for life. Galileo thought his headgear, with the telescope fixed over one eye and making its wearer look like a misaligned unicorn, would net him the reward.

Determining latitude is easy for any sailor who can pick out the North Star, but finding longitude escaped the citizens of the seventeenth century because it required a precise knowledge of time. That's based on a simple principle: Say you set your clock before sailing west from Greenwich. Say when the sun hits its apex, the clock reads five hours past noon. Because you know the Earth rotates 15 degrees per hour—completing the 360 in twenty-four hours—you know you're 75 degrees west of London. Easy peasy. That only works, though, if you have a clock that can keep accurate time, which nobody sailing the high seas then did. Clocks were complex mechanical devices, ill-suited to sea voyages. The rolling oceans messed with the pendulums that kept time on land. Salt air messed with everything. And explorers increasingly interested in crossing the oceans had a dangerously limited understanding of where they were.

Galileo's solution lay with the incessant eclipses of what he called Jupiter's four moons (and what astronomers now call Jupiter's Galilean moons, having discovered seventy-five more). In 1612, Galileo realized he could use the movement of the moons, which he charted, as a sort of astronomical clock. Over the next century, this became a common way of determining longitude on land. But it didn't work on the ocean. The telescopes of Galileo's age had tiny fields of view, and finding and tracking things that are at least 365 million miles away was virtually impossible on a ship bouncing along the waves. Galileo thought that by wearing the telescope on his face, a navigator could counteract that pitching and rolling more easily, and keep his focus on Jupiter's moons.

The Spanish were unimpressed, and no one ever claimed the king's ducats. Then, in 1714, the British Parliament passed the Longitude Act, offering £20,000 for a solution. After decades of trial and error, working class clockmaker John Harrison presented the marine chronometer, a time-keeping device that could withstand the rigors of the seafaring life and keep its practitioners on the maps they were filling in as they went along.

That successful iteration of the longitude prize was what Tony Tether had in mind when he concocted the Grand Challenge, a way to motivate the Harrisons of the world as well as the Galileis. But Tether was building on a long tradition that included failures as well as successes. Through the modern age, monarchs, philanthropists, scientific groups, and promoters of various ilks had enticed people to solve their problems with dollars, ducats, francs, rubles, lire, pounds, and more. Napoleon asked for the best way to preserve food, to better keep his troops fed. The solution involved boiling food and storing it in champagne bottles, an early variation of canning. (It also kept Napoleon's enemies fed at the Battle of Waterloo.) The British wanted a fertilizer as potent as the guano they spent a fortune importing from Peru. They never got it. The Duke of Northumberland offered cash for a self-righting lifeboat. That one worked out. Innovation prizes got the world's thinkers to try stamping out cholera, boll weevils, cattle plague, and the vineyard-destroying aphid that caused the Great French Wine Blight in the 1860s. None of those succeeded. In 1995, the Ansari X Prize pledged $10 million to the first nongovernmental organization to send a manned spacecraft into space twice in two weeks. Microsoft cofounder Paul Allen and aerospace engineer Burt Rutan claimed it in 2004.

The closest analog to Tether's Grand Challenge came on Thanksgiving Day in 1895, when six motorcars lined up for America's first car race, a fifty-two-mile course through the snow-covered streets of Chicago. H. H. Kohlsaat, the publisher of the *Chicago Times-Herald*, loved the new automotive trend and thought the race would sell newspapers. The cars of the era were hand-built to order, but the engines, transmissions, and other bits that made them run were easy enough to acquire. The trick was putting them all together so that the machine ran as reliably and quickly as possible. After ten hours in freezing temperatures (which by one account left one competitor comatose), just two cars crossed the finish line. First place went to J. Frank Duryea, driving the vehicle he had built with his brother Charles. The Duryeas went on to pioneer the mass manufacturing of cars, helping to spark America's enduring love affair with the automobile. More than a century later, Tony Tether just hoped his race would go as well.

"I myself have absolutely no idea what's going to happen this Saturday," Tether told his audience. March of 2004 had at last arrived, and the DARPA director was addressing the scientists, engineers, and executives who had come from the country's defense contractors and research universities to the Marriott Hotel in Anaheim, California, across the street from Disneyland. At that year's DARPATech conference, program managers from the Pentagon's innovation department showed off a handheld translating gadget called the Phraselator, for soldiers in foreign territory. They talked about microphotonic tech for detecting

biological weapons. They updated the crowd on plans to launch satel-
lites from hypersonic airplanes and ran through everything else (the
nonclassified stuff, anyway) on which DARPA would spend $3 billion
that year. But Tether was most excited about the Grand Challenge, set
for that weekend.

Tether spoke at DARPATech on Tuesday, March 9; the qualifying
round of DARPA's race for autonomous vehicles had begun the day
before, at the California Speedway in nearby San Bernardino County,
paring down the field before the main event in the Mojave Desert that
weekend. "I do know this," Tether continued. "No matter what happens, the DARPA Challenge will represent a huge success. Why? It has
sparked interest in science and engineering to a level possibly not seen
since the early days of the Apollo space program. Even if no team wins,
thousands of people have been inspired to develop technology that ultimately will save the lives of untold numbers of US soldiers."

Tether was right, at least about the huge numbers of Americans gal-
vanized to take on a problem that had stumped the world's best-staffed
and best-funded universities and companies. Even at other DARPA
events, race coordinator Jose Negron found himself surrounded by
wannabe competitors who had questions about rules and regulations.
More than one hundred teams had signed up, presenting the problem of
separating the crack engineers from the crackpots. Negron had started
the cull by demanding a technical paper from every team, laying out its
vehicle design and technological approach. It wasn't proof anyone could
actually build or fund what they planned, let alone that it would work,
but it was an easy way to weed out the truly clueless. Of the eighty-six
completed applications, the DARPA crew outright rejected about half
and deemed a quarter "completely acceptable," granting them access
to the qualifying round. The remaining teams would get a visit from a
DARPA official to determine if they'd be invited as well.

"Mostly they didn't have anything," said Doug Gage, a DARPA program manager who paused his work on LifeLog—an electronic data-

base that would record every moment of a person's life—for a weeklong series of site visits that covered both coasts. It wasn't until his third or fourth meeting that he saw a vehicle with wheels. But the enthusiasm was palpable. Tether himself trekked to Missouri to meet with one contender, only to find himself in a face-off with the man's wife. "Are you the guys that have got my husband trying to mortgage this house?" she demanded. Her husband suggested they have their discussion at the diner down the road.

Ultimately, DARPA invited twenty-five teams to the Qualification, Inspection, and Demonstration event at the California Speedway in Fontana, set to run from March 8 to 12, before the race itself on the 13th. Instead of NASCAR racers roaring around the oval track, those in attendance found the action focused on the grassy infield. Negron's team had marked off a 1.4-mile course with hay bales and concrete jersey barriers, littering it with obstacles mimicking the desert route waiting for the finalists: plastic barrels, cattle guards, a GPS-blocking overpass, parked cars, moguls, and more. As at the Petersen Museum gathering a year earlier, the crowd was an unlikely mix of geeks and government types, but this time with more military bigwigs in medal-specked uniforms, auto industry insiders curious to know if this was for real, and flocks of reporters and TV crews.

Those journalists had no better symbol of the wackiness of this government affair than the gangly grad student bent over ninety degrees as he pushed a supposedly autonomous motorcycle through the crowd. By the time of the qualifying round, Anthony Levandowski had his talking points straight. He told reporters a robotic two-wheeler could roam territory too rough for bigger vehicles. On the civilian side, he said his chassis control system could help motorcycle manufacturers detect imminent crashes and deploy airbags. He also lowered expectations by downplaying his shot at winning, noting that Ghostrider didn't even hold enough fuel to finish the course. But the fact that he couldn't win the race outright didn't mean he intended on walking away a loser.

He already had credit for trying something that no one else had even thought up. Now he needed to prove he could do it. And as he walked down pit row, he got his first look at the competition.

In an area usually the province of gasoline-soaked racing pros, hordes of sweaty, stressed nerds in baggy polo shirts worked on their Jeeps and dune buggies and ATVs. Every vehicle looked like it had crashed into a RadioShack and come out the other side wrapped in a labyrinth of cables, computers, cameras, radars, laser scanners, antennas, and whatever else its creators hoped would help it reach the finish line. Defense contractor Oshkosh had joined with the Ohio State University and Italy's University of Parma to make a robot from a lime-green, fourteen-ton, six-wheeled military truck. Team Ensco, a small Southern California defense company, had built what looked like a bathtub on wheels. The Golem Group, a band of engineers with ties to UCLA, had made its entry from an off-roading Ford F-150 pickup, using $50,000 its leader had won on *Jeopardy!*

Standing out, even in this colorful crowd, was Axion Racing, a team led by Melanie Dumas, an engineer who had sat at the Petersen Museum thinking this feat was impossible. She and her teammates had spent the past year turning a Jeep Cherokee into a robot, spending their time at their day jobs thinking about it and their free hours (many of which had once been sleeping hours) doing it. Their funding came from Bill Kehaly, an investor in a startup that imported bottled water from the tiny Micronesian island of Kosrae, who saw the Grand Challenge as a great marketing opportunity and named the robotified Jeep the Spirit of Kosrae. Eager to make the effort stand out, Kehaly had bought an aftermarket horn that made Spirit honk like the General Lee from *The Dukes of Hazzard*. He had also hired identical twins Julie and Shawnie Costello. In their late twenties, they were a few years into a joint acting career that included recurring roles as the "Juggy Twins" on Comedy Central's *The Man Show*. Kehaly brought them in for a photo shoot and produced what surely has to be the world's first autonomous vehicle calendar, featuring the blond twins. And he paid them to attend the race itself.

For the past year, these teams had been working in isolation. Apart from a chance encounter while testing in the Mojave or a bit of info gleaned from press coverage, nobody knew what the others were bringing to the Grand Challenge. Even the DARPA people had never seen most of the robots in action. Now though, they were all here, and each team would be asked to show what they had built and what it could do. And if any of them had forgotten who the odds-on favorite to win the Grand Challenge was, they just had to look up, where a skywriter hired by Boeing was spelling out: "Go Red Team."

The Carnegie Mellon powerhouse had been invited to the qualifier, of course, and arrived in force with its rehabilitated robot. As its convoy pulled into the raceway parking lot, several dozen students and associated others streamed out, clad in matching red shirts and hats. But it wasn't numbers, or even Red Whittaker's experience, that had made the team the clear-cut front-runner. It was the efforts of the team's core members, the dozen or so zealots who had spent months in that rust-wrecked steel mill, then more months sweating and cursing in the desert. Those who had wrecked Sandstorm in their drive to perfect it, then resurrected the robot in under a hundred hours. Those who had dropped everything else they had going on to meet DARPA's challenge.

Their competitors ranged in size, but many had witnessed a similar phenomenon. On Day 1, their initial meetings—in kitchens, garages, classrooms—had been packed. The idea of building a self-driving car had an undeniable appeal. On the checklist of *The Future*, it ranked with jet packs and lunar colonies. And what more fun way or good reason to deliver on the dream than at the behest of the American government, with a million-dollar prize? But it hadn't taken anyone long to realize how hard the Grand Challenge was. It wasn't the washouts, or the railroad tracks, or whatever other perils Sal Fish had set in store for these pioneers. The work that had seemed like it should be simple—just getting a car to drive itself in a straight line—was fiendishly hard. It was expensive, and budgets that ranged from tight to nonexistent didn't include room for paid labor. With the exception of Red Whittaker and his disciples, most everyone assumed failure. They just hoped for the least embarrassing version thereof. So in the yearlong run-up to race day, many people who had signed up fell away, minimizing their involvement or quitting altogether. Yet on every team, a core group remained. Tony Tether's idea for a race had captured them, but not for the reasons he pitched to the world. None of them saw the million-dollar purse as a way to get rich. Saving the lives of soldiers in the Middle East was an abstract ideal. And none really thought they were launching a new industry.

Even before picking up her keyboard, Melanie Dumas of the Axion team had known how difficult this challenge was. Plus, she was well into a successful career. During the run-up to the Challenge, she'd gone from working on software for voice recognition in tanks to the Predator drone. Chris Urmson had been headed for a solid life in robotics with the likes of NASA or Caterpillar before Whittaker pulled him out of the Chilean desert. Anthony Levandowski hadn't known where he was headed, but his knack for engineering and enterprise offered him a world's worth of options. None of them needed the Grand Challenge. But it grabbed them all the same, somehow convincing them it was worth a year without social contact, worth long sleepless nights, worth the risk of being run over by a robot. So when their cars stalled and crashed and rampaged, they didn't yield. They thought about how to find the problem and how to fix it. And when that solution inevitably broke something else, they went through it again. Learn, rework, repeat. Again and again and again, until time was up.

These were the people Levandowski watched as he walked past team after team, racing to make last-minute fixes and tweaks before they were ordered into the field. Now all that was left was seeing what they could do.

A few hours into the qualifying round of the Grand Challenge, Tony Tether wondered if his contest would look even more miserable than the snowy, coma-inducing *Times-Herald* race of 1895. DARPA's original rules said that to qualify for the final, a vehicle would have to navigate the full course twice. Tether and Jose Negron quickly realized that wouldn't leave them much of a race. Of the eight vehicles called to the starting line on the first day of qualifying (the order was set at random), none completed the course. The first few didn't manage to move at all. Some deferred their runs, realizing they had more work to do. Others brought their robots to the starting line, only to agonize as they sat there, motionless. TV cameras rolling, Tether looked at Negron. "Is anybody going to move today?"

DARPA's less than rigorous culling process and the short timeline of the Challenge meant many of the teams invited to the qualifying round simply weren't ready. But there was no extending this deadline. They'd come to the desert with what they had. Over the next few days, the DARPA officials filled out a score card that read like a mash-up of the Book of Revelation and Dr. Seuss: "hit blue van," "hit red car," "went left out chute," "went right out of chute," "endless circles," "crushed fence at bleachers," "serpentine spasms," "froze," "flipped on straightaway at 30 mph," "failed to move," "hit minivan," "hit fence," "trampled the cattle guard," "convulsions, smoke."

Carnegie Mellon's Red Team would have loved some extra time, too, to fully fix their robot after its ghastly crash less than a week prior. But when Sandstorm got to make its debut run at the end of the second day of qualifying, it validated its status as the favorite. It pulled out of the gate and rumbled over the small dirt mounds that made up the first leg of the course. As it rolled along at a few miles per hour, following the GPS coordinates DARPA provided, it swung its new gimbal back and forth, giving the laser scanner inside a full view of the path ahead. The Humvee went left, maneuvered past a pair of parked cars, made another left, rolled through the sand pit and a shallow ditch, passed another car, and went over a cattle guard. It grazed a hay bale here and there, but also clocked a brisk 19 mph. On the home stretch, Sandstorm headed for a gap formed by two concrete highway barriers. Then a pair of DARPA officials in fluorescent vests rolled a large metal sheet into the space between the barriers, blocking the way. The robot had just a few seconds to spot the blockade and safely stop.

From the sidelines, the students whom Red Whittaker had pushed to their limits could only pray Sandstorm would do what they'd designed it to do. That the cameras and the laser scanners in the gimbal Yu Kato had calibrated would detect the barrier. That the software Chris Urmson and Kevin Peterson had written would understand that their data meant something was now in the vehicle's way, that it couldn't drive over or around that something, and that it had to stop. They prayed that decision would make its way from the massive red electronics box Matt Johnson-Roberson had packed with computers, through the thick black winding cables, to the emergency brake system Nick Miller had crafted from junkyard parts. And finally, that the machine would do what a human's foot does so naturally: hit the brakes.

As the robot approached the barrier, even Whittaker, watching through binoculars and normally so cavalier, expressed his doubts. "This is a little tricky," he said. Then, Sandstorm slowed. A few yards before reaching the barrier, with a slight screech of the brakes, it stopped completely. It had passed the test. The crowd whooped. "We've now made history, ladies and gentlemen," the announcer called out over the PA system. Whittaker grinned and leaned over to kiss his wife. In that day's journal entry, he promised Sandstorm would run "at hotter speeds" for its second run, to ensure it secured pole position for the final. And he rewarded his troops. "Grades will be good," he wrote.

Glad the Grand Challenge had its first finalist, Negron and Tether scrambled to ensure Carnegie Mellon would have some competition. They set up a practice area where teams could adjust their machines to the particulars of the course, loosened the rules to give everyone as many runs as they could fit in, and reassured the media everything was going great. Gradually, the field expanded as teams muddled through. Axion's Jeep faltered about a third of the way through the course on each of its first two runs, but reached the finish line on Thursday, the last day of qualifying. So did five other teams. The day before the race, DARPA announced those seven teams would advance to the final, along with eight more that hadn't crossed the finish line but had performed well enough.

For those watching the qualifying, Anthony Levandowski's Blue Team had seemed a long shot for that latter group. Even in a weak field, the motorcycle's struggles stood out. Levandowski had put so much time into just getting the thing to stand up that he'd had little time to actually make it navigate. And it still didn't stay up too reliably. The day Sandstorm first completed the course, Levandowski deferred his initial chance at a run. Over the next few days, he put his bot into the field several times, only to see it careen left and right before smashing itself into the ground. One day, the motorcycle made three attempts in a half hour, each time flopping over. But when DARPA published its list of fifteen finalists the day before the Grand Challenge, there the Blue Team was, in the last slot.

Neither Tether nor Negron expected the motorcycle to get anywhere near the finish line, but they recognized that Levandowski's wild idea didn't just embrace the open nature of their Grand Challenge. It encapsulated it. He had even made it work, kind of. And the DARPA guys knew that as much as they were hosting a robotics competition, they were also putting on a show. "It was such a PR attraction that we had to bring it to the desert and see what it did," Negron said. "Tony and the group felt like it merited moving forward, to at least try."

Perhaps unintentionally, Levandowski had made it to the final with the kind of hack, or shortcut, he always admired. The sheer nuttiness of the motorcycle ended up lowering the bar for what it had to accomplish. The robot couldn't do much, but DARPA couldn't deny it.

Levandowski loaded Ghostrider into his pickup and headed for the Mojave.

———

The journey from the California Speedway to Barstow started with the interstate, I-15, that Tony Tether had hoped to shut down for his imagined race from Disneyland to the Las Vegas Strip. Cutting north and away from Los Angeles through the Inland Empire, the freeway climbed into the chaparral-covered hills, running parallel to old Route 66 and between the Angeles and San Bernardino National Forests. After one final looping bend, the eight lanes of freeway spat travelers into the Mojave Desert. From here, the asphalt ran dead northeast, past the brownish yellow towns of Hesperia and Victorville. After forty-three miles of flat shrubby landscape, it reached Barstow. Originally settled in the mid-1800s, the town had picked up in population when miners flocked to Southern California in the 1860s and '70s. The railroads that followed the money crossed paths here. By 2004, Barstow was home to some twenty thousand people, a Marine Corps base, the second largest meteorite ever found in the US, and not much else. Another ten-minute drive out of downtown, heading south on State Route 247, led to the Slash X Cafe.

On a normal day, the Slash X's parking lot was filled with dirt bikes and ATVs, their riders inside downing beers and steaks so big they hung off both sides of the plate. Sal Fish called it "a bloody shit-kicking cowboy-type place out in the middle of nowhere," and had picked it as the starting point of the course DARPA had recruited him to design. With license plates on the walls and old trucker hats pinned to the ceiling, its vibe didn't match that of the defense agency that had birthed the internet and the Predator drone. (The khaki-clad wildlife biologists brought in to keep the robots away from the endangered desert tortoise, accustomed to dusty lifestyles, didn't seem to mind.) But it had a big parking lot, land to spare, and easy access into the Mojave.

Jose Negron had organized a barbecue at the Slash X for Friday evening, after everyone had made the trip out from the qualifying round at the Speedway and before the race the next morning. But while the bigwigs held forth on the American spirit of innovation and the tortoise biologists and race marshals and judges chowed down, the men and women whose talents they were tapping were forgoing the festivities. Instead, they were in the staging area, an expanse of dirt DARPA had cleared of junked cars and old tires. With just hours left before their shot at a million bucks and one-of-a-kind bragging rights, they sat hunched over laptops in trailers and tents, adjusting their software. They crawled into, over, and under their robots, fiddling with sensors and checking their wiring. They had given up enough sleep over the past thirteen months, since that meeting in the Petersen Museum, that one more night didn't matter.

That final night ended at four in the morning, when Negron handed each team a CD-ROM detailing the course. This list of 2,586 latitude and longitude coordinates, each about a hundred yards from the one before it, would lead the robots from the Slash X to the finish line in Primm, 142 miles away.

While most teams simply uploaded the coordinates into their vehicle's navigation system, Red Whittaker's mapping squad sat in its trailer and went to work on the task it had spent months training for. The route the software spat out said it would take Sandstorm thirteen hours to reach the finish line. But the goal was to finish in ten hours or less, and Whittaker wasn't about to settle. Neither was his team. Sitting shoulder to shoulder in front of a bank of computers, each of a dozen Red Team members studied the waypoints for the section of route he'd been assigned. The team compared them with their database of satellite maps, evaluating the difficulty of the terrain and calculating where the Humvee would need to run slow and, more importantly, where it could gun it. They entered a specific speed for every bit of the course, and loaded all the data into the computers whirring away in the huge electronics box. Their planning was aggressive, but this was a race. If Sandstorm could pull it off, it would be hard to beat.

At 6:30 a.m. on Saturday, March 13, 2004, with the sun just starting to peek over the hills to the east, the Grand Challenge was finally starting. Safety monitors and reporters were in place along the route. Hundreds of people filled the bleachers. A helicopter whirred overhead, carrying cameras and the electronic kill switch that could stop any robot in its tracks. DARPA crew members sat in the chase vehicles that would follow each entrant through the course, with the same gear. The biologists had run a final tortoise sweep, and the first six vehicles sat in a semicircle, each in its own starting gate. They would enter the course at five-minute intervals, followed by the rest of the field. The Red Team, based on its impressive performance in qualifying, would go first. Chris Urmson prepared Sandstorm for its final journey, turning on its various systems, waiting for the green "It's All Good" lamp to signal that the emergency brake was primed, and flipping a silver switch from manual to auto.

The national anthem played and Tony Tether took his seat next to a four-star army general who had showed up without warning. Then the race announcer went to work. "The command from the tower is to move," he called out. The flag dropped. Yellow light flashing and siren sounding, Carnegie Mellon's Humvee rolled out of the gate. "Ladies and gentlemen, Sandstorm! Autonomous vehicles, traversing the desert with the goal of keeping our young military personnel out of harm's way," the announcer went on. "Booyah!"

Negron and Sal Fish had made the opening moves of the course deliberately simple. The GPS coordinates would take each vehicle out of the starting gate, through a couple of easy turns, and past the grandstand. Then it was through a couple of cattle gates and onto Powerline Road, a dirt path barely distinguishable from the brown land around it. As long as a vehicle could stick to the road and handle any rocks or tumbleweeds it encountered, the first few miles would be easy going.

As Sandstorm turned the first corner and started to disappear into the open desert, its laser scanner detected a clear path ahead, and its detailed map data said it was time to hit the throttle. Sandstorm's wheels spun, kicked up dust, and carried it away at more than 30 mph.

For Chris Urmson and his teammates, the moment resonated. This was the first time they had let their robot out of sight and out of their control. There was nothing more they could do. No more testing, no more fixing. Sandstorm would complete the course, or it wouldn't. And its first big test was just a few miles ahead. The trick to getting up and over Daggett Ridge was mastering the switchbacks, hairpin turns so tight that following a GPS path alone could easily send a vehicle tumbling off the path and hundreds of feet down. So could a misaligned sensor, or any number of software glitches. If you could clear that hurdle, however, it was back to flat ground and mostly clear roads, nearly smooth sailing all the way to Primm.

Over the next twenty minutes, three more vehicles left the gate, following Sandstorm's trail. It looked like Tether was going to get a proper race, even after the shaky showings in the qualifying round. Then the problems started.

Sixth off the line was Axion Racing. Over the past year, Melanie Dumas's skepticism and reluctance had transformed into swelling optimism. She had seen her team's Jeep drive in this kind of terrain, and drive well. She even thought that with a bit of luck, the Spirit of Kosrae might outrun Carnegie Mellon's Sandstorm. When the flag waved, the Jeep pulled out of the chute and made the first turn smoothly. But as it approached the first narrow gate, it turned again. All the way around. There was no obvious reason for the about-face. Maybe the sensors had deemed the opening too tight. Perhaps something else had acted up. It didn't matter. As the Jeep drove back to the starting line, sending its chase vehicle backward like a linebacker, DARPA hit its emergency shutoff. Axion's Grand Challenge was over in a matter of seconds. Dumas was devastated.

Next up was the University of Louisiana's six-wheeled Cajunbot. It smacked a wall on the way out of the chute, knocking itself out of contention. It was followed by Ensco's bathtub of a bot. As the flag waved, it stood frozen for a few seconds, rolled forward, stopped, then started again. It drifted to the left, where the edge of the road sloped upward, tilting to one side before moving back to flat ground. Then it went left again, this time too far. It flipped over and landed on its side, one thousand feet into a 142-mile course. The whole run lasted one minute and six seconds.

A group of students from Palos Verdes High School had spent the night before the race scrambling to fix the steering controls for their vehicle. At the last moment, they settled on a solution they hoped would work, with no time to test it. Their prayer went unanswered. Their entry, Doom Buggy, never turned at all. It rolled out in a straight line and after fifty yards hit a concrete barrier.

SciAutonics I, led by an engineer who'd worked on Germany's autonomous driving efforts in the 1980s, saw its ATV wander off the trail, never to return. The University of Florida's Cimar strayed off course half a mile in and got tangled in a wire fence. Terramax, the fourteen-ton, lime-green, six-wheeled military truck, went 1.2 miles before getting stuck between a pair of small bushes that its sensors mistook for immovable obstacles. Tired of watching it lurch back and forth like a driver trying to escape an impossibly tight parallel parking space, Tony Tether ordered the kill.

More than one person had compared the field of homemade robots to something out of a *Mad Max* movie. Now they had the carnage to match.

Ignorant of the slaughter happening at the Slash X, Sandstorm carried on ahead, driving smoothly down the dirt road. But the race officials in the chase vehicle soon noticed troubling behavior. Nine minutes into its run, the Humvee approached a pair of fence posts about twelve feet apart, at a road crossing. Its laser sensor picked up the obstacles, and Sandstorm should have been able to slide through with a foot to spare on either side. But it hung to the left and smacked head on into one post, which didn't stand a chance against a Humvee moving at speed. Neither did its compatriot at the other side of the intersection, which met the same fate a few seconds later.

Ninety seconds after the double whammy, Sandstorm stopped, possibly because of an inadvertent signal from DARPA's emergency stop system. It got moving again after five seconds, but when it strayed into yet another fence post forty-four yards up the road, it didn't have the speed to blast through it. The Humvee wasn't programmed to give up, though, and it pushed against the barrier for nearly two minutes before the post gave way. Sandstorm went the next thirteen minutes without incident, then smacked into a boulder on the left side of the road. Its recovery software sent the robot too far back to the right, then corrected.

This "off-nominal behavior," as Chris Urmson and his teammates put it in a postmortem report, was not disqualifying. DARPA had no rules against destroying a bit of property on the way to Primm. But some flaw in the Red Team's system was leading their vehicle astray, and there was no fixing it as the robot headed for the narrow road that would lead it up and over Daggett Ridge.

The robot worked its way through the first few turns of the road without incident. Then it headed into one hairpin a little bit wide. As it turned the corner, it cut toward the inside, going just a touch too far and pushing its left wheels off the road. Sandstorm kept going, straddling the embankment and lifting the Humvee just enough that its wheels couldn't get the traction to move forward. The computers didn't know the vehicle was beached, so they stayed on the throttle. For three and a half minutes, Sandstorm spun its wheels, until the rubber melted off the front tires. Watching the smoke pour off the robot, a DARPA crew member hit the emergency stop. The wheels were going so fast, the clamping of the brakes snapped the half shafts that connected the transaxle to each wheel. That last bit of damage didn't matter, though. Just 7.4 miles into the Grand Challenge, Sandstorm was dead.

Back at the Slash X, Chris Urmson and his teammates had no idea how Sandstorm was doing until radio reports started to trickle in. As with most disasters, the first bits of news were hard to piece together: The robot was on fire, or its tire had come off, or it was stuck on a rock. It soon became clear, though, that the robot was finished. All that work—the freezing nights at the steel mill, the brutally hot days in the Nevada desert, the endless practicing with the mapping software—and the team had made it just 5 percent of the way to the goal they had always refused to consider out of reach. They had come in as the favorites, the powerhouse to be reckoned with. And they had come up short. Some cried. Others let the exhaustion finally, fully hit them. *At least we get to go home now*, Matt Johnson-Roberson thought.

By 9 a.m., Tony Tether knew he wouldn't get to grin for the cameras as he handed someone an oversized check. The other vehicles that had looked good at the start met their own fates within a few miles: blocked by a rock, stuck on an embankment, stalled going up a hill. But one vehicle had yet to leave the start line: Anthony Levandowski's motorcycle.

Ghostrider's miserable showing in the qualifying had made clear that it wasn't going far. Its small gas tank guaranteed it wouldn't finish the course. But Levandowski didn't have to reach Primm to come away a winner. If the motorcycle could go a few miles, it would prove that he had the engineering chops to back up the gutsiest idea DARPA's Grand Challenge had engendered, and prove a lot of naysayers wrong: the Berkeley faculty who had chuckled at his idea; the many, many potential sponsors who had turned him down; the established robotics community as a whole, whose recipe for autonomous driving Levandowski had ignored. He was the underdog, the kid with the crazy idea, the leader of the team that made the world's first self-driving motorcycle. Ghostrider would show the world what he could create.

Levandowski rolled his robot cycle into the starting gate, set it to run, and passed it off to a DARPA official who would hold it upright until it was launched. The flag dropped. Ghostrider started to roll forward. The DARPA man took his hand off.

The motorcycle fell to the ground.

Levandowski ran to his machine, flapping his long arms in distress. In the heat of the competition, and after two days without sleep, he had forgotten to flip the switch to turn on the stabilizer system his team had spent a year developing. DARPA's rules were clear: Once the vehicle started the race, it couldn't accept any human help. Ghostrider wasn't about to raise itself from the ground. It was done, dead. The bike had crashed somewhere between six hundred and eight hundred times in the past year. None hurt like this one, because this was the one time it was supposed to work, when everyone was watching. "Good try, guys," the race announcer said. "Anthony, good effort."

Tony Tether got into the helicopter watching over the race and headed to Primm, where a gaggle of reporters was waiting to see the first robots cross the finish line. One asked him how it was going. "Well, it's over," Tether said. "You know, the farthest car went 7.4 miles, caught fire, blah blah blah. It's over."

The DARPA Grand Challenge was indeed over, and the headlines would mock it as an overhyped science project. "Foiled: Darpa Bots All Fall Down," *WIRED* sassed. "Nobody won. Nobody even came close," CNN wrote. "The reality of the event did not come close to meeting the hype surrounding it," said tech site the *Register*. On that count, *Popular Science* blamed the feds: "If DARPA was plainly guilty of anything, it was not managing inflated expectations. Instead of billing this inaugural Grand Challenge as a not-ready-for-prime-time field test to calibrate what was needed for future efforts, race manager Negron, in the months leading up to the checkered flag, continued to predict a victor."

DARPA didn't have that victor, but an embarrassed Tether chose to focus on the positive. Watching these kids in the desert, competing but coming together, trading stories and strategies, he didn't worry about the steel carcasses and burned-up tires they had left behind. His great race had proven the motivation to create autonomous vehicles was out there, and it had found the talent with the drive to make it happen. He wasn't about to give up. A reporter asked him what he would do now.

"We'll do it again," Tether said. "And this time, the prize will be $2 million."

— 5 —

Vision Quest

THE NOTION OF MACHINES THAT THINK FOR THEMSELVES stretches back into the mist. The Greeks told of Talos, the bronze humanoid who hurled boulders at Jason and the Argonauts to keep them from his island. Taoist and Buddhist scriptures contained stories of automatons nearly indistinguishable from the humans who created them. And in 1955, the idea took on the imprimatur of academia, when computer scientist John McCarthy coined the term "artificial intelligence." McCarthy played a prime role in turning the pursuit into a proper discipline, helping found the Massachusetts Institute of Technology's AI Lab. In 1965, he launched the Stanford Artificial Intelligence Lab, better known as SAIL.

Based in a low, semicircular building tucked among the hills of Northern California, where the door signs were written in J. R. R. Tolkien's Elvish language, SAIL became the center of the early, heady days of AI research. Between games of volleyball, its men and women sparked a surge of accomplishments that seemed to put manufactured human intelligence in reach. They taught computers to play checkers and chess. They explored computer vision and speech recognition systems and designed a prosthetic arm that could sort a pile of blocks by size.

The lab produced eighteen winners of the Turing Award, the computer science equivalent of the Nobel Prize. Its alumni carried on their work at places like Microsoft, Sun Microsystems, Apple, Xerox PARC, and Cisco. Most of the funding for this whirlwind of progress came from a single source: Between 1963 and 1973, SAIL researchers spent nearly $10 million of DARPA's money. By 1973, the agency charged with out-innovating the world was supporting two-thirds of the lab's 128 scientists.

One of those pioneers was Hans Moravec, the Austrian-born computer scientist who worked on the Stanford Cart. While his colleagues cheered on their chess programs in battles against those developed by Carnegie Mellon and the Soviet Union, Moravec watched his wheeled robot struggle to navigate clutter-filled rooms in one-meter hops punctuated by fifteen-minute thinking breaks. Its failure rate—the frequency with which it would whack into whatever sat in its way—sat at a stubborn 25 percent. "Good enough for my thesis, perhaps," Moravec wrote, "but not good enough for a robot to do complex tasks that would, at the minimum, require it to cross rooms many times."

Watching his and others' robots flounder, Moravec identified a paradox he laid out in his 1988 book *Mind Children: The Future of Robot and Human Intelligence.* "It has become clear that it is comparatively easy to make computers exhibit adult-level performance in solving problems on intelligence tests or playing checkers, and difficult or impossible to give them the skills of a one-year-old when it comes to perception and mobility," he wrote. Humans, after all, were the result of a billion years of survival-driven evolution favoring the ability to understand and move through one's surroundings. The work that went into mastering chess might seem more complex than snapping one's fingers, but it was actually just a newer trick.

Making a capable, let alone graceful, robot, Moravec argued, was so difficult because it relied on consciously recreating ways of seeing, thinking, and moving that are deeply ingrained in the brains of humans and animals. In 1997, IBM's Deep Blue chess program beat world champion Gary Kasparov in just nineteen moves. In 2004, the convulsing funeral pyre that was the DARPA Grand Challenge confirmed the enduring difficulty of making a machine move through the natural world. But in the years after he laid down Moravec's Paradox, the evolutionary gap started to close.

In the 1990s, several new approaches to robotics had combined to promise a breakthrough. One was probabilistic robotics, which was based on the acknowledgment of a stubborn fact: When dealing with robots, there is no certainty. Putting a real machine in the real world means accepting that its sensors might not perceive everything properly, that its motors might not be perfectly tuned, that its perception of its surroundings and ability to move through them might therefore be flawed. The trick, then, was to build robots with a statistical understanding of that uncertainty. Moravec had experimented with this approach in the mid-eighties, with good results. But the main push to build the mathematical models that would make the unknowns known came from a German researcher by the name of Sebastian Thrun.

Born in West Germany in 1967, Thrun was just twelve years old when he re-created his favorite arcade video game on his NorthStar Horizon home computer. After earning his PhD at the University of Bonn in 1995, he moved to America. Carnegie Mellon had spotted him as a rising star and brought him in as a computer research scientist.

While in Pittsburgh, Thrun continued working with two colleagues from Bonn, Dieter Fox and Wolfram Burgard. In 1997, the trio's robot, RHINO, acted as a docent at a local museum. It looked like a trash can with a computer stuck on its lid. It used laser range finders, sonar sensors, and an infrared camera to navigate, and the probabilistic techniques its creators developed to determine where it was in space and how to move from one exhibit to another without hitting anything. It was Thrun's idea to set RHINO loose in the museum. More than just a dynamic test environment, he wanted a place where the public could interact with the robot—a novel experience for virtually everyone in 1997—and see what computer science could do. He and RHINO both got a lesson in humanity. Kids blocked the robot's path, hoping to make it honk its horn. One group of cleverer, meaner adults tried to send it toward a flight of stairs. During its weeklong test, RHINO stayed on solid ground, and gave short tours to more than two thousand visitors.

Thrun and his Bonn colleagues followed with Minerva, an upgraded robot that spent two weeks in the summer of 1998 guiding visitors between exhibits at the Smithsonian's National Museum of American History. Next came NurseBot, a friendly looking machine designed to escort elderly nursing home residents to physiotherapy appointments, open medicine bottles, and the like. When it was time for them to write a paper, Fox and Burgard, still in Germany, would send their thoughts and notes to Thrun, in Pittsburgh. The next morning, they'd wake to find he'd finished the thing, in excellent English.

The ability to write a paper in a night, in whatever language, was but one of the qualities that won Thrun his ticket to the United States. Before long, he accumulated a score of 140 on the H-index, a measure based on how many papers one publishes and how often those are cited by others. That put him among the most influential computer scientists of all time. While at CMU, Thrun kept cranking out papers and robots, including an enhanced walker that could park itself, then find its user when summoned by remote control. He worked with Red Whittaker on Groundhog, a robot that mapped old mines.

After a few years in the US, Thrun had lost some of his German accent and most of his hair. He retained his youthful looks, though, perhaps thanks to his seemingly constant good cheer. His bright blue eyes and often mischievous smile worked together, as if to ask whether you were with him on whatever adventure he'd stumbled into. He loved talking to reporters, and was always ready to explain how whatever he was working on could apply to real people's real lives. And he never seemed to take himself too seriously. In one photo from a ski trip, he's lying in front of the group in the snow, posing like a swimsuit model.

In 2003, as Red Whittaker's Red Team was preparing for the Grand Challenge, Thrun was in the midst of leaving Pittsburgh for Palo Alto. Stanford had hired him and asked him to reopen SAIL, which John McCarthy had shut down in 1980 after research slowed and DARPA funding dried up amid what's now called the first "AI winter." Thrun was just thirty-six years old, but he came to the university as a tenured professor. He was hired, in part, for his work in the budding field of machine learning. In this branch of artificial intelligence, the computer gets lessons, not laws. Say you want to make a computer recognize a cat. You don't type in rules about pointy ears and whiskers. You show it photo after photo of cats—thousands of them, every one labeled "cat"—until it learns what to look for. The idea had been around since the 1950s, but only started to gain ground with the invention of new techniques in the eighties.

Machine learning is what took Carnegie Mellon's Dean Pomerleau and Todd Jochem on their 1997 No Hands Across America drive. Instead of labeled photos of cats, Pomerleau drove his car down the highway, then fed his system images of road markings paired with his steering direction at each moment. When the lines went straight ahead, the wheel didn't turn, and the computer learned the car should go straight ahead. When the lines curved to the right, the wheel turned to the right, and the computer learned to follow suit. It worked well enough to get them from Pittsburgh to San Diego, and even into a meet and greet with car fanatic Jay Leno.

Thrun didn't take part in the Grand Challenge, but he was in the Mojave the day of the Slash X massacre. DARPA had invited him to show off his latest project, a Segway-based device that used lasers to map the world around it and determine its own location within that map. Even without the invitation, it was the kind of event he couldn't miss. And he was disappointed by the efforts of his fellow roboticists.

"Pretty much everybody I know walked away thinking, 'My god, let's hope that's not all there is to it,'" Thrun said. As he drove back to Palo Alto, he decided to try for himself.

DARPA waited until June of 2004 to officially announce that, yes, it would be holding a repeat of the Grand Challenge, and that the winner would take home $2 million. Tony Tether soon saw that the debacle of the race hadn't deterred a new field of contestants from joining up: Nearly half the teams from the first go reenlisted, joined by more than one hundred sixty newcomers. This edition would follow the same rhythm as the first go-round, with a slightly higher bar for would-be racers. Along with a technical paper, every team would have to submit a video of its technology in action and allow DARPA to come see it in person. Tether's agency would then invite the forty most viable teams to a qualifying round for the first week of October 2005. Half would advance to the Grand Challenge itself, on October 8. This time, the course would be a series of loops around Primm, Nevada.

In July, Thrun began his effort. His first teammate would be Mike Montemerlo, a postdoc researcher who'd followed him from Carnegie Mellon to Stanford. The son of a NASA program manager, Montemerlo earned his bachelor's, master's, and PhD from CMU; his thesis advisors were Thrun and Red Whittaker. Quiet and unassuming, the software savant avoided the spotlight in which Thrun thrived. Joined by David Stavens, a computer vision specialist and one of Thrun's first PhD students at Stanford, they started to consider how to approach the problem of sending a robot on a solo mission through the desert.

Watching the implosion of the first Grand Challenge, Thrun had formed a simple diagnosis: The vehicles didn't see anything. The hardest problem in mobile robotics hadn't changed since the days of the Stanford Cart. The newly created team started by reviewing the literature around computer vision, investigating different laser sensors and cameras, and exploring how it could all be put together. And they got their vehicle when they first heard from Cedric Dupont.

Dupont was an engineer with the Electronics Research Laboratory, the Palo Alto research and development center Volkswagen had established in 1998 to dip a ladle into Silicon Valley's pool of talent and ideas. The lab's few dozen engineers were investigating things like haptic touchscreens, electronically tintable glass, and handwriting and voice recognition. When Dupont heard that Stanford was joining the Grand

Challenge, he secured permission to get the lab involved in a supporting role that played to its strengths. He presented Thrun with a whop-
per of an offer: Volkswagen would build Stanford a car, taking on all
the work of translating a computer's commands to the working of the
throttle, brake, and steering wheel.

Thrun appreciated the value of this contribution because he understood what many teams had missed the first time around, especially
those that had arrived with the kind of custom-designed vehicles that
convulsed and smoked. The Grand Challenge was not a hardware problem. With the right tires, any Jeep, Humvee, or ATV could have sur-
vived the 142 miles from Barstow to Primm, if it had had the software
to replicate the human driver. (This was easier to see in hindsight, as
those teams had to create their vehicles before seeing the reality of the
course.) VW was offering Thrun expert engineers to handle the necessary but uninteresting mechanical work, freeing him to focus on the
crux of the problem. He happily accepted.

Designing robots for touring museums offered valuable experience,
but sending one galumphing through the desert at 30 mph was new for
Thrun and his colleagues. To get a handle on the problem, they decided
to start by building a prototype. As with drawing a sketch before putting oil to canvas, the point was not perfection. It was to create a road
map that revealed which parts of the project would require the most
attention. Quick and dirty, this was the sort of effort that demanded
manpower. So Thrun added CS294: Projects in Artificial Intelligence
to the Stanford Computer Science Department's course listings. One
day in the last week of September, a group of master's, PhD, and undergraduate students gathered in a basement classroom of the Computer
Science Building that Microsoft's Bill Gates had funded.

Thrun had a reputation as a dynamic lecturer, the sort of teacher who
could connect the theory of computer science to its real-world applications, and do it without seeming to have prepared at all. But this was a
different sort of course. There would be no syllabus, Thrun told the students. No weekly readings, no lectures. They would spend the next two

months building an autonomous racing robot. Their final exam would come on December 1, when they would head to the Slash X Cafe and see if they could get the thing to drive the first mile of the course of the original Grand Challenge.

In early October, Dupont and the Volkswagen crew delivered their vehicle. They had modified a small Touareg SUV by hacking into the electronic systems that worked the throttle and brakes, sticking a chain drive onto the steering column, and installing an aluminum arm to move the gearshift. They added skid plates to protect the car's underbelly in rocky terrain and rerouted the air-conditioning to cool the six Pentium M blade computers living in the trunk. The team dubbed the robot Stanley.

The twenty students in Thrun's class, meanwhile, divided into groups and started to build the computer and perception systems they believed would matter most. At every step, Thrun emphasized hardware simplicity. Instead of a complex gimbal that could point the laser sensor in any direction, like the one Carnegie Mellon had used, they put five laser pointers on the roof, each angled in a slightly different direction. They would rely on clever software to make up the difference.

Progress came quickly. By November, Stanley could credibly be called an autonomous robot, even if its software was full of bugs, its obstacle detection skills were on par with Mr. Magoo's, and it needed a human in the driver's seat to supervise. On the dirt roads near their quonset hut (where Thrun had struck a space-sharing deal with Stanford's solar car project), Stanley would occasionally mow down a trash can it was meant to avoid or shoot off in the wrong direction, forcing whoever was behind the wheel to hit the red plunger button that killed the autonomous system. But it could follow GPS waypoints and mostly tell what was road and what wasn't.

Then came December, and with it the deadline for putting Stanley to the test on the original Grand Challenge course. The team piled into a fleet of rented SUVs, drove the six hours to Barstow, grabbed dinner at one of the few restaurants in town, and caught a showing of *National Treasure* at the local, second-run movie theater. The next day, they got up early and gathered in the parking lot of the Slash X. Shivering in the wind, they checked that all the robot's sensors were properly hooked up, turned off the car, took the key out of the ignition, and hooked up the chain drive that controlled the steering column. They turned on the computer systems in the trunk, along with the DARPA-mandated siren and light on the roof. They powered up the GPS, laser scanners, and camera, then the software systems that controlled them, along with the software that confirmed that the rest of the software came online correctly. They loaded in the same route file that DARPA had distributed the morning of the Grand Challenge and slotted in the pin that connected the metal arm on the gearshift to the motor that moved it.

With Mike Montemerlo watching the software and the rest of the team ready to follow in a motorcade of SUVs, Thrun got behind the wheel and flicked the three switches that gave the computer control of the throttle, brake, and steering. Now in charge, Stanley pulled away from the parking lot that had witnessed the deaths of so many robots nine months earlier. It weaved back and forth and moved very slowly, but soon hit the one-mile mark, satisfying Thrun's goal.

Stanley kept going, following Powerline Road up into the hills where the switchbacks started. After 7.4 miles, Stanley reached the turn where Carnegie Mellon's Sandstorm had met its fiery end. Ignorant of this history, the robot went right by. Thrun, holding his left hand by the wheel and the right over the buttons that would disconnect the autonomous system, grinned at the triumph. Then, a mile later, the car hit a steep downhill stretch where it should have tapped the brakes, but didn't. As Stanley accelerated down the mountain, Thrun brought the robot to a stop. It wasn't enough for a million dollars, but it was better than anyone had managed the first time around. And this robot—made in a few months—was but a sketch of the masterpiece he had in mind. Before heading back north, Thrun picked up a souvenir: a chunk of wood, a remnant of one of the fence posts Sandstorm had smashed on its failed run.

With ten months until race day and eager to avoid a "too many coders spoil the bot" situation, Thrun traded his army for a SWAT team. Its core would be him, Montemerlo, David Stavens, a handful of the top students from the class, and a Volkswagen engineer named Sven Strohband, who took over the hardware work when Cedric Dupont left the automaker's lab. To fund the effort, Thrun struck sponsorship deals with venture capital firm Mohr Davidow Ventures, Android (which had created a new mobile operating system and was about to be acquired by Google), Red Bull, and Intel, which loaned some researchers on a part-time basis.

The rebooted team quickly fell into a rhythm, developing the robot's software at Stanford and making one or two trips a month to Barstow to test the robot on the original race course. While Montemerlo wiped most of the code they'd programmed into the robot—rewriting it from scratch was the only way to ensure it met his exacting standards—Thrun and his small crew turned their attention to making Stanley see.

To reach the finish line, Stanley would have to find the road and spot any obstacles in its way. For the latter, the five laser scanners on the roof were the obvious choice. If there was a boulder, a bush, a tree, or whatever else in their path, they would spot it every time. The trouble was, the car's constantly changing pitch on the rough road made the scanners "see" things that didn't exist. Over and over again during test runs in the desert, Stanley would perform just fine for a few miles before suddenly swerving off the unobstructed road and justifying the team's policy of wearing white racing helmets while aboard.

The conundrum led Thrun back to his fields of probabilistic robotics and machine learning. Giving Stanley an understanding of when its lasers were right and when they were crying wolf, though, demanded a trove of reliable data. One way to get it would be chaining a grad student to a computer and having her pore over camera and laser data, labeling every last pixel as "drivable" or "not drivable." Thrun cooked up a faster, simpler, more humane approach. He got behind the wheel and drove himself, telling the car to consider everywhere it went as free and clear. If the lasers thought they saw a tree but Thrun went right through it, Stanley learned to dismiss that data as a false positive. Before this machine learning exercise, nearly 13 percent of the "obstacles" the lasers "detected" didn't exist. A few hours' worth of training dropped that rate to .02 percent—one in fifty thousand.

That solved the lasers' reliability issue, but not their other weakness—myopia. To stay ahead of whatever reinvigorated robot Red Whittaker was surely cooking up, Thrun estimated Stanley should be able to drive at 35 mph. The problem was that the lasers could see only twenty meters out. If Stanley was going 35 mph, by the time it spotted a boulder in its path, it would be too late to avoid a crash. Like a human driver on a foggy night, it had to choose between speed and safety.

The Carnegie Mellon team had solved this problem with its deep-benched mapping crew, manually going through satellite data and determining how fast the car should go at any given point. Thrun was determined to do it with software. The car's camera could see all the way to the horizon, but fifty years of artificial intelligence research had failed to deliver a camera system that could identify all the road surfaces a desert crossing would include: brown and yellow dirt, pavement, metal cattle guards, and more, all in lighting conditions that would change throughout the day. Then Bob Davies, one of the Intel researchers helping out, suggested combining the lasers' reliability with the cameras' range.

They called the result adaptive vision. When the lasers verified that the upcoming twenty meters were clear, the camera would compare what it saw in that same stretch with what it saw farther out. If the images didn't match, it would slow down. If the ground eighty meters ahead looked just like what was right in front of the vehicle, Stanley could accelerate, knowing the way ahead was clear. This spin on machine learning, where the laser provided the training data, was a conceptual coup, not a mathematical breakthrough. And it was just the kind of thing Tony Tether had hoped his Grand Challenge would deliver.

Nearly as important as these advances was the team's ability to stay on task. Thrun considered it his job to shield his comrades from what he called "too-many-ideas syndrome," whose symptoms involved wasting precious time trying out new sensors or techniques that would at best make Stanley a slightly better driver. Instead, he used their meetings to keep everyone on deadline, and to make sure every choice they made was backed up by hard data. Then, they'd go test it in the real world.

Over the first half of 2005, Thrun and his teammates learned all about the indignities of life in the Mojave Desert, those sweltering days bookended by breakfasts at the Ramada Inn and dinners at Sizzler. Flocks of birds would confuse the sensors, sending the robot zigzagging down the road. One day, they got Stanley stuck in the mud up to its doors. After the team in the chase car came back from the nearest town with shovels, boots, and two-by-fours, they spent four hours in 110-degree weather digging the robot out. At dinner that night, Montemerlo—not an especially big person—ate a twenty-three-ounce steak. Another time, during a shift change, he passed his helmet to his teammate Hendrick Dahlkamp, who didn't notice the cricket inside until he had it on his head. Montemerlo, with a mop of curly hair he rarely bothered to cut, hadn't noticed it at all. But bad food, hard labor, and six-legged invasions didn't stop them from transforming Stanley into a remarkable robot.

On that first drive in December 2004, Thrun had thought the robot steered "like a drunken squirrel." In early July, Stanley ran the original Grand Challenge course start to finish, 142 miles, without a misstep. Even that didn't impress the Stanford researchers and their comrades as much as the time they went for a drive after a rainstorm had flooded parts of the road. As the car went along, it kicked up so much water, the wipers struggled to keep the windshield clear. The humans couldn't see a thing, but they didn't stop the test. Stanley's roof-mounted sensors had a clear view, and the robot was rolling along at a steady pace. "We're relying now only on software?" Strohband, the VW engineer, asked from the front passenger seat, not quite masking his concern. The Stanford researchers answered with giddy laughs, tinged with nerves. A moment later, the water slid away and the windshield was clear again. Stanley carried on, unaware that for a brief moment, it had surpassed its creators.

Any sense of celebration, though, was muted by stubborn imperfection. One day in the fall of 2005, *New York Times* tech reporter John Markoff visited the team in the Arizona desert and went for a ride with Thrun and Montemerlo. All went well for a dozen miles or so, until Stanley misread an overhanging branch as an obstacle and swerved off the path, landing in a thornbush.

Anthony Levandowski took off his sneakers and his socks and rolled his khaki pants above his knees. He checked that his cell phone wasn't in his pocket as he stepped into the water. The twenty-five-year-old engineer had seen his robotic motorcycle crash many times, but watching Ghostrider take a hard left turn off the road, climb a small embankment, and dive into a water retention pond at Berkeley's Richmond Field Station was about as bad as it got. Especially when the bubbles rising to the surface confirmed that the bike was flooded. Crouching over, Levandowski hooked one end of a tow strap onto the bike and the other onto the back of his pickup truck. As one of his teammates drove forward, Levandowski held the bike upright and guided it back onto land. He turned it on its side, watching the water pour out.

A motorcycle, Levandowski had by now accepted, was not a good way to win the Grand Challenge. The difficulty of keeping the thing on its two measly wheels wasn't worth the payoff in maneuverability. But the embarrassment of watching his motorcycle fall over like Wile E. Coyote running off a cliff—hope of victory suspended, then dashed—left him with unfinished business. He had promised a self-driving mo-

torcycle. The flop outside the Slash X Cafe had made him look like a
fraud. He had to prove he could make the concept work.

The two-wheeled approach had worked in one sense: Just about
every news article concerning the Grand Challenge mentioned the
Blue Team and Ghostrider. The media christened him the charismatic
underdog, and Levandowski leaned into the persona. The second
time around, he didn't have to cold-call potential sponsors. He rebuilt
his team, bringing back some veterans and adding fresh blood. Their
burrito-fueled work schedule was as crazy as ever. They'd start late in
the morning, then go until 2 or 3 a.m. Often, they would come back the
next day to realize Levandowski hadn't gone to bed at all.

The bike, with its 90cc engine sounding like a lawn mower, never
stopped crashing. But this time around, Levandowski wouldn't let a single stumble doom their effort. His team added a screw jack to each side
of the bike. If Ghostrider tumbled to the left or right, the arm on that
side would slowly extend, lifting the bike back onto its wheels. They improved the motorcycle's sight to the extent that they could, rigging up a
three-camera setup to distinguish the flat terrain from the bumpy stuff,
and, in theory, keep the bike away from obstacles. For serious testing,
Levandowski and his crew went to tiny Lake Winnemucca, northeast of
Reno. Over hundreds of test runs (every single one ending in the bike
falling down, its antenna whipping back and forth), the Blue Team sent
the motorcycle over wooden beams, between trees, and through traffic
cone slaloms to gauge its durability and agility. They jogged along behind it, like parents teaching a kid to ride a bike.

They rebuilt the motorcycle's engine twice and fiddled with its sensors and software endlessly. And they created a checklist, to ensure that
everything that needed turning on would be turned on when race day
finally arrived.

For Red Whittaker, DARPA's double-or-nothing offer was a no-brainer.
Now with an experienced crew and a robot that had proven it could

handle at least some of the desert, he gathered an astounding $3 million in funding and built a one-hundred-strong army. "Two hundred percent effectiveness, or you've got some work to do," the old marine would bark at his recruits.

They started by giving Sandstorm a makeover. Working hours just as intense as the first year, they inspected and cleaned every component, dumped sensors they had not used, simplified the cabling, and installed air-conditioning to cool the electronics. They made the gimbal that held the laser range finder smaller and more agile, and filed a patent for the design. They added accelerometers to measure the motion of the Humvee's suspension and chassis, to improve its ability to choose the right speed. The mapping team redoubled its efforts, gathering data for every square inch of land within thirty miles of Primm. One grad student created a system that, every second, calculated 750,000 S-curves the robot could execute if it needed to avoid a last-minute obstacle. In November 2004, Chris Urmson finished the PhD he had delayed to join the Red Team, defending a dissertation based largely on his work with Sandstorm, called *Navigation Regimes for Off-Road Autonomy*. By that point, the headless Humvee was back in Nevada, pounding through the desert. In a contest that hinged on endurance and reliability, there was no such thing as too much testing.

The team members who stayed in Pittsburgh turned their attention to the second part of Whittaker's comeback plan. The first race had made clear that nothing could guarantee you wouldn't end up surrounded by smoke. Now the Red Team would show up with two robots. Sandstorm would be joined by a shiny new Humvee, a gift from its manufacturer, AM General. H1ghlander (the name a nod to Carnegie Mellon's Scottish founder) would get all the sensors and software Sandstorm had. And it would get to keep its roof. Because each team could enter only one robot, Whittaker split his crew in two: Red Team and Red Team Too. The line between them was superficial; they followed one game plan. By the end of 2004, H1ghlander was driving itself around the old steel mill site.

As October 2005 steadily approached, the Red Team banged and typed away until their twin vehicles could average two hundred hours of driving between computing or hardware failures, an order of magnitude more than they'd need to complete the race. Yet for all their effort, the Carnegie Mellon outfit didn't quite exorcise the demons of the first race. During testing in Nevada, with a week left before the qualifying round, H1ghlander sidetracked to the left, hit the raised edge of the road, and flipped into the air. As it landed on its side, the Red Team members once again scrapped plans to get some sleep before the big day and scrambled to revive their robot. And once again, in a matter of days they had their champion up and running again. Almost as good as new.

In 2005, the Red Team didn't have the same monopoly on experience it carried into the first Challenge. Then, Carnegie Mellon had housed a substantial percentage of the people in the country who had a serious grasp of how to build an autonomous robot. The 2004 race changed that. Everyone involved, no matter how badly they had failed, now understood what it would take to win. And when qualifying rolled around again in October of 2005, they proved they could execute, too.

For the second Grand Challenge, Tony Tether's crew had again turned the infield of the California Speedway in Fontana into a simulacrum of the course Sal Fish had designed in the desert. The 2.5-mile track included parked cars, an artificial hill, a tunnel to block GPS, stretches of rough road, speed bumps, straightaways, and a narrow "mountain pass" demarcated by cones. If the opening hours of the 2004 qualifying round had proved just how unready most teams were for the race—many incapable even of putting a robot into the field—the start of the 2005 trials showed what some experience and an extra eighteen months could do. The first vehicle to take the field was a Nissan SUV called Xboxx, put forward by a group of machinists from Colorado. It completed the entire course. Over the rest of the day, six more robots followed suit. Among them was the Spirit of Kosrae, the Jeep that Melanie Dumas's crew had converted into an autonomous robot. Cheered on by the Axion Racing twins, Spirit ran the qualifying course without drama and earned itself another shot at first place.

Less conspicuous was a team called Intelligent Vehicle Safety Technologies 1, running a computer-covered Ford F-250 pickup truck called Desert Tortoise. Like so many unimaginatively named business ventures run by people who don't say much about where they're from, the team was a front. Its leader was Jim McBride, a Ford engineer who had volunteered at the original Challenge as a race marshal, figuring he might see some promising tech that his employer could work into upcoming models. Instead he came away thinking, "I could do a hell of a lot better." He made his case to his bosses in Dearborn, who gave him a small team. Theirs was a fact-finding mission: They were to scope out new ways of making Ford cars safer and more attractive to customers, and figure out if this autonomous driving thing was anything more than insanity. Above all, they were to avoid embarrassing the company. "It just wouldn't play well for a major auto company to plant a large-sized truck in the side of a building or drive off the edge of a cliff," McBride said. The automaker was ready to take off its socks and maybe even dip a toe in the water. It wasn't ready to swan dive in, and it certainly wasn't ready to be seen belly flopping.

McBride's crew, though, didn't have to worry about drawing too much attention. Most of the people pressed up against the concrete highway barriers that separated the human and robotic realms focused on the new front runner that had emerged.

Sebastian Thrun's team had spent much of 2005 at Volkswagen's expansive proving ground in the Sonoran desert outside Phoenix, Arizona. With access to the top-flight facility, they refined their robot until it could run hundreds of miles at a time. Thrun had even set up a testing division of his team, which created a one-hundred-page document detailing ordeals designed to expose any and all of Stanley's weaknesses. By the time it reached the Speedway, the Stanford robot—freshly waxed by the Volkswagen contingent—could run the entire qualifying course perfectly. It set a steady pace, its laser scanners and camera together analyzing each bit of the course before proceeding. In between runs, the team relaxed in their garage. The VW guys sipped beers. Mike Montemerlo gave a tour to his friend Chris Urmson—they'd shared an office at CMU. Thrun, friendly and outgoing as ever, chatted with reporters. He boasted that his robot was doing so well that the DARPA crew tasked with resetting the course, after each robot inevitably hit something or other, had dubbed Stanford "Team Boring."

Red Whittaker, though, didn't give way easily. His two teams' Humvees stomped a hay bale here and there, but posted times nobody would match. At the end of the weeklong qualifying round, DARPA gave pole position to H1ghlander, Whittaker's first team. Stanley would be second out of the gate, followed by Sandstorm, Whittaker's second team. Melanie Dumas's Jeep would go fourth.

Anthony Levandowski's motorcycle had a harder time. Agility didn't make up for the fact that its tri-camera setup just couldn't see as well as the sensing systems its competitors used. On its first run in the qualifying round, the motorcycle plowed directly into a black, billboard-shaped piece of metal that marked one side of a faux cattle gate. On its second and third runs, it fared little better. The obstacles it didn't hit, it avoided by going entirely around—and off the course. Sometimes it popped itself back up after falling, sometimes it didn't.

On its fourth and final attempt, Ghostrider got nearly halfway through the course. Then it clipped the side of a tunnel and fell onto its left side. The kickstand sprang into action, tipping it up—and over onto its right side. The other jack tried to push the bike back onto its wheels, once, twice, three times, never getting it all the way upright. Ghostrider collapsed into the dirt, defeated. There would be no place in the finals for Levandowski's kooky idea this year. DARPA had a full field of twenty-three real contenders.

Levandowski had taken the motorcycle idea as far as he could, and was ready to let it go. He always had an eye out for the next opportunity, and he had already spotted one. It sat on the roof of the green Toyota Tundra pickup truck sitting in the garage bay directly opposite his.

When Tony Tether said he wanted the Grand Challenge to bring in the sort of person who had never worked with DARPA, he was looking for Dave Hall. In his early fifties, Hall wasn't exactly some kid in his parents' garage. He was, though, a master of looking at complex problems and finding novel solutions.

Born in Boston in 1951, Hall came from a line of engineers. His father helped design nuclear reactors in Illinois and Connecticut. When he was four years old, Hall built a Heathkit amplifier. A few years later, he rigged a motor to a minibike for his little brother Bruce to ride. As a teenager, he made a turbocharger for his BMW 2002, then took the car racing. When he wasn't building something, he was taking it apart, eager to reveal its inner workings. Hall was never much of a student, making a point of walking the Case Western University campus without any books. After graduation, he followed a self-guided course of study, focusing on whatever interesting problem strayed into his orbit. He started a machine shop outside Boston, bidding on contracts to build things like laser tables for defense contractor Raytheon and high-pressure balloons for surgeons pioneering angioplasty. Hall liked the job, making his own hours and working with his hands. At thirty-one years old, he moved to the San Francisco Bay Area. He had ended up there a few years earlier after a cross-country bike ride and liked the weather and the people. Setting up in San Jose, he launched a company called Velodyne, whose speakers won rave reviews for their ability to pump out bass without distortion.

What made Hall an exceptional engineer was his understanding of every part of whatever system he worked on. By the time he was ready to try to build something, he knew every element intimately. Where a traditional company would pass a design from one department to the other—machining, electronics, manufacturing, and so on, with the associated losses in translation and costs in overhead—Hall did it all himself.

Riding the popularity of the CD, the DVD, and home theater systems, Hall's business boomed, his speakers selling for thousands of dollars apiece. Then in the late nineties and early two thousands, the rise of Chinese manufacturing wiped out many of his American subcontractors, and pushed Hall to start importing more than he was building. "Instead of manufacturing and inventing, I was a rug merchant," he said. He did not relish the role.

It was around this time that Hall heard about an interesting contest for robots. Anybody could enter *BattleBots*, and if your machine successfully smashed, cleaved, or incinerated its competition, you could win a bit of money. It meant getting on TV, which Hall thought was neat, as well as a good way to promote his company. Working with his brother Bruce, who'd joined Velodyne to help run the business side of things, he built Drillzilla. The 350-pound robot looked like a VCR and hacked its way to second place in the world championships.

By 2003, *BattleBots* had ended, and the Hall brothers were looking for something new. Then they heard about the Grand Challenge. This sounded like much more fun than fighting robots, and it might be a way to redirect the struggling company. If the military was serious about building autonomous vehicles, Dave figured he could find a way to cash in.

Tony Tether thought all the ingredients were there—the computers, the sensors, the hardware—he just needed someone to put it all together. Dave wasn't so sure. "By definition, whatever the government says is wrong," he said. "They say it's the secret sauce, so it's got to be something other than that." He bet an autonomous vehicle would need a new kind of sensor, a better way to see the world.

Dave and Bruce formed Team Digital Auto Drive, aka Team DAD. (Their slogan was "Are we there yet?") Instead of using the basic laser sensors most teams relied on, they modified Dave's forest-green Toyota Tundra pickup with a stereoscopic camera system. Like everything Dave invented, it was elegant in its simplicity. The two cameras would watch the road, and the work of picking out obstacles and translating that intel into driving decisions was done on a single circuit board, which fit into the truck's roof rack.

But the camera software produced a lot of false positives. The car was constantly stopping or veering off course because it confused a puddle or a shadow for an obstacle. So on the day of the first Grand Challenge in 2004, the brothers took a gamble, turning off the system and relying on GPS alone. (Thus supporting Sebastian Thrun's theory that most of the Grand Challenge robots were blind.) Six miles in, the move was paying off. Then DARPA paused the Tundra, so race officials could deal with the crippled Sandstorm on the narrow paths of Daggett Ridge. As it stopped, one of the Tundra's front tires rolled up against a football-sized rock. When they started the truck back up again, it didn't have the speed to get over the rock. Sitting in the Slash X parking lot, Dave and Bruce learned they had finished in third place, and felt not a little bit robbed.

Dave spent the first year of the runup to the 2005 Grand Challenge working on the software that governed his camera system, trying to stamp out the false positive problem. Six months before the follow-up race, he watched his vehicle flunk a private test run, coming out of a turn and smacking into a barrel that was outside the camera's field of view. He realized his truck would need another two sets of cameras to see everything that could knock it out of the competition, and a way to seam all their images together. He had already done the fun part—the design—and now faced the prospect of months of tedious building. The depressing notion led his mind back to a conversation he'd had at the Slash X with Jim McBride, the Ford engineer who had volunteered as a safety marshal at the race (and who was quietly running Ford's 2005 entry). McBride had said Hall might want to look into Lidar.

Lidar by this point was a well-established technology. It was first developed in the early 1960s; in 1971 the Apollo 15 astronauts used the laser scanning technique to map the surface of the Moon. Mostly, it was a survey tool, used for mapping and meteorological research. Red Whittaker first used a Lidar scanner on an autonomous vehicle in the late 1980s, with his Terregator robot. It was a logical choice for navigation, firing pulses of infrared light and measuring how long they took to return after hitting whatever object they encountered first. It provided precise, trustworthy data—*object, 6.87 meters dead ahead!*—and could see in the dark as well as the light. But going into the second Grand Challenge, most teams had to settle for a laser scanner made by a German company called SICK. It cost a few thousand dollars and was intended for far simpler uses than traversing the desert, like detecting when a truck had pulled into a garage, or someone's hand got too close to a robotic arm on an assembly line. Because it used just a few laser beams, the picture it yielded was the computer equivalent of looking at the world through a mail slot. Stringing together half a dozen or more Lidars could help, but brought up the messy question of how to combine their data into a cohesive picture. McBride suggested to Hall that if he could get more laser beams into one unit, he might get a better view of the world.

Hall dumped his camera system and started gnawing on this new bone. He read up on Lidar and bought hobby-grade components to play with. He used an oscilloscope to detect the return pulse of the Lidar, and his own circuit boards to control the thing. He did the math to figure out how many laser beams might be enough to guide his truck through the desert. Instead of one, he would use sixty-four. To provide a full field of view and avoid the pain that comes with stitching together results from different units, he made the whole contraption spin around—no small thing when you're talking about a device stuffed with wires and computer chips. When he needed help, he pulled his engineers away from their day jobs, sacrificing Velodyne's speaker R&D for what had become a new passion.

The process took six months, and by the end Hall had built what he and Bruce called "the big wheel of death." Weighing about three hundred pounds, his Lidar consisted of eight groups of eight laser beams.

The whole thing spun around, collecting sixty-four thousand data points about its surroundings every second, in 360 degrees. The visualization software Hall created to render its data in a human-friendly format had an advantage over the SICK Lidar that was stunning. In terms of detail and resolution, it was like going from a game of *Pong* to *Call of Duty.*

Dave finished building it just two weeks before the qualifying event for the second Grand Challenge. While Bruce drove them to the California Speedway, Dave sat shotgun with his computer in his lap, writing code. Their truck cruised through the qualifying round, earning a place in the final, set for October 8.

Primm, Nevada, didn't have much going for it. Home to about a thousand people, it sat on land once owned by a gas station proprietor and bootlegger known as Whiskey Pete. The town existed mostly because it was the closest bit of the vice-happy Silver State to Southern California, just off the I-15 freeway. Course designer Sal Fish had often used Primm as a staging ground for off-road races, and DARPA picked Buffalo Bill's Resort and Casino as the home base for the 2005 Grand Challenge. The fifteen-story building had 1,242 rooms and a pool shaped like a buffalo, all of it surrounded by a monstrous roller coaster once among the fastest and steepest in the world. What attracted Fish, however, was the sprawling parking lot, which made up a sizable chunk of "downtown" Primm and abutted the Mojave Desert.

For the big day, DARPA set up a series of large white tents, one to serve as a race officials' command center, one to hold the media, and one especially big one for participants and members of the public to hang out in when they wanted to get out of the sun. After all, there was no point sitting in the grandstands once the vehicles had left the start line.

DARPA brought a two-hundred-person crew to the new Grand Challenge, this time led by Tony Tether's chief of staff, Ron Kurjanowicz. (Jose Negron had left the agency in February of 2005, part of

DARPA's constant and natural churn.) His command center tent looked like a cheaper version of a NASA mission control center. Sitting in folding chairs behind folding tables, headset-wearing DARPA officials tracked each vehicle on brick-thick laptops and large screens, making sure they knew where everyone was and what was going on. The rest of the staff would be outside, working as environmental monitors, tow truck operators, or law enforcement. Some followed the robots in chase vehicles, holding the remote kill switches that would freeze any that went rogue.

The grandstands were bigger and fuller than the first time around, and studded with tech world celebrities. Apple cofounder Steve Wozniak explored the pit lanes on his Segway, wearing a bicycle helmet. Larry Page, the Google cofounder who'd dreamed of autonomous cars since he was a kid, showed up.

While many teams had spent the final hours before the 2004 Challenge scrambling to rewrite code and adjust hardware, the scene on the eve of the second race was calm. Most of the engineers had learned their lessons the hard way, but this time their work was done and their vehicles were ready. They spent the final night enjoying an open bar and free barbecue, mingling and swapping war stories with their fellow racers.

At four in the morning on race day, Red Whittaker stepped out of Carnegie Mellon's trailer command center. He marched to the wooden booth where Ron Kurjanowicz was handing out the CDs with the waypoints for the race. The men shook hands and Whittaker strode back to his squad. Moments later, CMU's mapping team was poring over each little bit of the route. They gave their two Humvees detailed paths for moving from one waypoint to the next, including the maximum speed they could safely drive at any given moment. And they programmed in a last-minute strategy change: Instead of running H1ghlander and Sandstorm at a pace that balanced speed with safety, Whittaker would play tortoise and hare. H1ghlander would drive as fast as was reasonable, Sandstorm a few miles per hour slower. Whittaker knew Stanford's car had skills, but that it wasn't the speediest. If H1ghlander booked it the whole way, Stanley shouldn't be able to keep up. And in case the first Humvee failed, Sandstorm would be chugging along behind it, slow and steady.

As the sun crested over the mountains and streaked the desert sky with purple, with human-driven cars going by in the distance, H1ghlander and Sandstorm took their places at the starting line. Both vehicles were red, while Sebastian Thrun's blue Stanley stood between them. After the national anthem, Tony Tether gave the latest iteration of his speech about innovation and America, and at 6:40 in the morning, he waved a green flag. H1ghlander shot off into the desert. Five minutes later, Stanley left the gate, hesitant at first, then picking up speed. Another five minutes, and the flag waved for Sandstorm. The veteran robot rolled out calmly and got to work. The hunt was on.

Sal Fish's 132-mile course started and ended in Primm, taking the robots on a series of loops through the Mojave. Running on narrower roads and tighter turns than any vehicle had faced a year earlier, the bots would have to dodge dozens of utility poles and pass through three GPS-blocking tunnels. They'd drive on broken pavement, gravel utility roads, and off-road trails. The winner wouldn't necessarily be the first to cross the line, but whoever completed the course the fastest. (If officials paused a vehicle for any reason, that time wouldn't count toward its total.)

With the Carnegie Mellon and Stanford vehicles on the road, the rest of the field followed. As each vehicle left the chute, the crowd could see what a difference eighteen months had made. Of the twenty-three teams that made the final, all but two would surpass Sandstorm's original 7.4 miles. Nineteen would make it ten miles or more; seven would reach the halfway point. Still, the course took its toll. Caltech's "Alice" jumped a concrete barrier. Team Ensco suffered a blown tire. A bug in the Ford team's steering software revealed itself fourteen miles in, and a DARPA official hit the kill switch before the pickup smacked into a utility pole. Melanie Dumas's Spirit of Kosrae was halfway through the course when it hit a washout that knocked a wheel out of alignment. Pulling to the right, it went off the trail and into silty terrain. The

Jeep dug itself into the fine dirt and hung itself up on a rock. DARPA declared it dead.

Three hours into the race, Tether was confident enough in the lead vehicles' performances to invite a group of reporters to head into the desert. They piled into a bus and went to a cordoned-off lot that looked onto the course's halfway point. Ten minutes after they arrived, they spotted H1ghlander's headlights coming up over the horizon, accompanied by a cloud of dust. A human-driven chase car followed closely behind while a helicopter whirred overhead, filming the race from the air.

In normal circumstances, a Humvee driving at 20 mph down a dirt road wasn't anything special. But H1ghlander was driving so crisply, sticking to the dead center of the road and holding a steady speed, that it was easy to forget there was nobody inside. The Humvee was reading the road and making decisions and executing them all by itself. A few minutes later, the journalists saw Stanley chugging along in H1ghlander's wake, looking just as competent, if a little dusty. Another few minutes and Sandstorm came along.

"I have an instinctive urge to cheer the vehicles on as they whiz past," reporter Elizabeth Svoboda wrote in her live coverage for *Popular Science*'s website, "but remind myself there's no one inside to appreciate encouragement from the sidelines."

After watching the three would-be kings crossing the desert, the journalists climbed back into the bus. Just before their human driver pulled away, Svoboda looked out the window to catch another glimpse of H1ghlander as it climbed a small hill—and suddenly slowed. The Humvee stopped, then rolled backward, as if someone had snuck inside and thrown it into neutral. After a moment, it regained its grip and climbed the incline. Svoboda, despite herself, let out a small cheer. What she didn't know was that that H1ghlander's stumble wasn't a fluke. For reasons that would remain a mystery for more than a decade, H1ghlander was crippled. It would spend the rest of the race poking along at half its programmed speed. The potency Whittaker suspected it needed to clinch first place was gone. And Stanley was closing in from behind.

Over the next hour, Thrun's car ate into H1ghlander's lead. Multiple times, DARPA officials froze Stanley so it wouldn't attempt to pass the Humvee on a narrow stretch of road. In the big tent, the CMU and Stanford teams tracked their robots on a screen that showed the course and the racers, marked by green squares for the moving robots and red for those that were stopped. At one point, staring at the red square that denoted Stanley, Thrun decided the robot must be dead.

After every pause, though, Stanley would hit the throttle and close the gap. At the 102-mile mark, the balance swung. In an open stretch of the course, the officials acknowledged the inevitable, and halted H1ghlander. From nearly a football field away, Stanley's camera spotted the hunk of metal ahead. *Better slow down*, it told the motion planning system. *There's something up ahead that doesn't look like open road.* When the gap closed to about twenty-five meters, the laser scanners confirmed it: *Big boxy thing, dead ahead.* With a few seconds to spare, the chain drive that connected to Stanley's steering column went into action, swinging the car to the left, putting a few feet between itself and the Humvee. It passed H1ghlander and kept on going, unaware as ever of such a momentous occasion: taking the lead and becoming the first autonomous vehicle to pass another. When the news reached the main tent outside Buffalo Bill's, the Stanford team started whooping. Their robot was in the lead.

Sal Fish's final gauntlet was Beer Bottle Pass. The ten-foot-wide winding road, wedged between a wall of rock and a steep drop, was the sort of stretch that could spook a human driver. Robots don't suffer such emotions. Stanley steadily moved along the twisting dirt path. As the DARPA helicopter circled overhead, the VW hit the final, flat stretch of the race and blasted for the finish line. Behind it, H1ghlander, still moving slowly, navigated the mountain pass. Then Sandstorm cleared it, too.

That was little consolation to Whittaker when he heard the race announcer pick up his microphone: "I see a dust trail in the vicinity!" he called. "Ladies and gentlemen, here comes Stanley!" The crowd, pressed up against the fence nearest the finish line, hooted and hollered as the

VW, its blue paint hidden by a thick coat of dust and grime, approached the finish line. Six hours, fifty-three minutes, and eight seconds after the car had taken off, Tether waved the checkered flag, his great idea realized at last. "God," he thought. "We made it."

After dousing Thrun with two buckets of ice water, the Stanford team hoisted their leader and Mike Montemerlo onto their shoulders. Thrun had made a lot of robots do a lot of cool things, but this beat them all. "This is the biggest day of my life," he told himself.

Eleven minutes later, Sandstorm reached the finish, followed after a few minutes by H1ghlander. Two more vehicles would finish the course. A group of engineering students from Tulane University in New Orleans proved to be the Cinderella story. Despite minimal funding and losing nearly a month of work time when Hurricane Katrina ravaged their city a few weeks before race day, their Ford Escape—cheekily named Kat-5—finished the race just sixteen minutes after Sandstorm. And the military truck fielded by defense contractor Oshkosh, the colossus stumped by a pair of tumbleweeds in 2004, made it so far the day of the race that when the sun set, Tether decided to let it keep going the next morning. It crossed the finish line on October 9, missing the ten-hour cutoff but claiming a victory nonetheless.

At the awards ceremony the following morning, Red Whittaker spoke with pride. His robots and his team had done a remarkable thing. But anyone could see that he was stung by the loss, and he took the blame. If he hadn't held Sandstorm back, it might have beaten Stanley. Being first among losers was bad. Being second among winners was worse.

The Great Imposters

WHATEVER WOULD COME OF BUILDING ROBOTS THAT COULD drive across the desert, the success of the 2005 Grand Challenge brought Tony Tether's career as DARPA director back from the brink. Tether had been humbled by the slapstick failures outside the Slash X Cafe in 2004 and had dreaded a repeat performance. More concerning was that before the first Grand Challenge, the public had learned about Total Information Awareness, the intelligence gathering and compilation program Tether had started, led by John Poindexter of the Iran-Contra scandal, as a response to the September 11 terrorist attacks. The reaction was fierce. Under the headline "You Are a Suspect," a *New York Times* columnist declared the project an "assault on individual privacy." The American Civil Liberties Union called the program perhaps "the closest thing to a true 'Big Brother' program that has ever been seriously contemplated in the United States."

DARPA's track record had always bounced not just from success to failure, but from programs with obvious benefits (like the internet) to those that whose uses were more militaristic (like the Predator drone). Though Tether considered Total Information Awareness a vital tool, there was no denying it fell into the latter camp. Its logo—the pyramid-topping Eye of Providence beaming its light on the globe—helped seal its fate. Amid a civil liberties furor, the Senate voted unanimously to gut the program's funding.

The undeniable and high-profile success of the 2005 Grand Challenge did much to salvage DARPA's reputation. The headlines were all good, and the stories supported Tether's narrative that 2004 was not a failure, but a necessary, if shaky, step toward victory. "In a Grueling Desert Race, a Winner, but Not a Driver," trumpeted the *New York Times*. "A Big Finish with No One at the Wheel," said the *Washington Post*. CNN and the Discovery Channel covered the race. PBS's *Nova* produced an hourlong episode about it, starring Red Whittaker, Chris Urmson, Sebastian Thrun, and Anthony Levandowski.

Tether was ready to declare victory and walk away. But soon after the race, his colleagues got on his case. *We haven't done anything yet*, they said. Tether was reluctant to put his recently saved neck back on the line. But over Chinese food at his favorite DC restaurant, Peking Gourmet (the kind of place with notable patrons' photos on the wall, including Tether's), Ron Kurjanowicz and Tether's other lieutenants hammered at him until he admitted they were right. The Grand Challenges had shown driverless cars could handle the desert if they didn't have to face any moving objects. To be useful, they'd have to do much more. It was time to put them into the real world, or at least a facsimile of it. Over several months, the DARPA crew spun together a plan that would challenge the robots to navigate city streets, intersections, parking lots, and more, and to do it all while working around other moving vehicles. The robots would have to navigate four-way stops and make left turns into traffic. They'd be ordered to look for open spots in parking lots and change lanes to get around stuck cars. And they'd have to do it all while obeying every rule in the California driving code.

On May 1, 2006, Tony Tether announced that DARPA would host one more driverless vehicle competition. It would take place November 3, 2007, with a bump in prize money: $2 million for the winner, $1 million for the runner-up, and $500,000 for third place. It would be called the DARPA Urban Challenge.

———

As DARPA was dreaming up a city, the Carnegie Mellon crew had returned to the desert. With no idea another Challenge was coming, Red Whittaker had found a practical use for his team's software at Nevada's Tonopah Test Range. Halfway between Las Vegas and Reno, this sere expanse was truly in the middle of nowhere, and for good reason—it was also known as Area 52. Over half a century, the military had littered the site with unexploded ordnance, and now the government wanted a good idea of where all those devices might be.

Through his company, Redzone Robotics, Whittaker won a contract to map part of the site without subjecting a human to monotonous back-and-forth driving or the nuclear material that early experiments had splashed around the place. The team migrated H1ghlander's navigational capabilities into a 100-horsepower tractor and used it to tow a platform of mapping sensors. Over two months in the fall of 2006, Tugbot, as it was called, worked twelve-hour days, surveying eighteen hundred acres of dry lake bed. For Kevin Peterson, who'd done his first serious robotics work on the Grand Challenge and was now nearing completion of his PhD, it was evidence of how their machines could do real jobs, making humans' lives tangibly safer, easier, better. As Whittaker liked to say, they were putting a dent in the world. But when Peterson heard his Carnegie Mellon colleagues were gearing up for the Urban Challenge, with the chance to forge a heftier sledgehammer, he didn't hesitate to enlist.

For the two Grand Challenges, the Red Team's roster had consisted of undergrads, graduate students, and volunteers, all more or less equal, and led by Whittaker. That changed for the Urban Challenge. Now the roster swelled with more than a dozen Carnegie Mellon senior researchers, folks who had worked alongside Whittaker for decades, pushing the limits of what machines could do. The shift reflected the new nature of what DARPA was demanding. Traversing the desert was a matter of finding the road and dodging any serious obstacles, none of which moved. Navigating through a city, even a mock one, was orders of magnitude more complex. DARPA's list of demands—intersections, parking, and so on—oozed with juicy problems around perception, logic, and motion planning. Solving them demanded increased brainpower, and the brains capable of the job were suddenly interested in signing up.

The change in scenery caught the eye of another group that had stayed away from the desert Challenges. Two hundred miles northwest of Pittsburgh, General Motors research and development chief Larry Burns heard about the Urban Challenge. He had followed the Grand Challenges, and had declined requests from various universities for sponsorships. Robots that drove across the desert were a bit outside GM's wheelhouse. The "urban" bit of this competition changed that. Regardless of DARPA's military messaging, anyone could see that computers that could handle city driving would have a serious impact on the home front. Sebastian Thrun in particular pushed that narrative. Whenever he spoke to the press—which was, approximately, all the time—he went out of his way to talk about how the technology could boost the economy and stop the crashes that were killing more than forty thousand Americans every year. It was the kind of world General Motors could get behind, and wanted to stay ahead of. "We've got to be involved," Burns thought.

The right move, he figured, was to imitate Volkswagen's sponsorship of Stanford University's Stanley, providing financial and engineering help to a university team. He saw no better partner than Carnegie Mellon. It was a natural match. Since the late nineties, the automaker and the university had jointly run a collaborative research lab to investigate driving-related technologies like connected cars, computer vision, and human-vehicle interface design, based in Pittsburgh. So GM invited Red Whittaker to its Warren Technical Center outside Detroit. The modernist landmark was designed by Eero Saarinen and built in the 1950s, the symbol of GM at the peak of its prowess. But by the early 2000s, the company that had wowed the 1939 New York World's Fair with its vision of the future had ceded its reputation for quality and innovation to the competition, chiefly the Japanese and Korean auto-makers who were winning over more and more Americans. Joining the Urban Challenge could signal to the world that GM could still innovate, and might also help the automaker develop new features for its own vehicles. As Carnegie Mellon's primary sponsor, GM would contribute $2 million, and send two engineers to Pittsburgh for the duration of the Challenge.

Whittaker's team picked up more cash from DARPA itself. The twin tenets of the early races—that all were welcome and that no one got special treatment—had elevated unconventional entrants like Anthony Levandowski's motorcycle and birthed surprise successes like Tulane's Kat-5. But the Challenges created an aristocracy of teams with the connections and expertise to dominate the field. Now DARPA forged a class system to go with it.

Participants in the 2007 contest would fall into one of two "tracks." Track A was for the heavy hitters like Carnegie Mellon, Stanford, MIT, and Virginia Tech, plus defense contractors Oshkosh and Raytheon. DARPA would give each of these teams $1 million to fund their re-search and development efforts. Everyone else was still invited to sign up, but those relegated to Track B wouldn't get any kind of help at the outset (those that made it to qualifying would receive $50,000; reach-ing the finals was good for $100,000). DARPA no longer needed any-one and everyone. It just needed winners.

To go with new levels of funding and the participation of top tal-ent, Whittaker got the full support of his university. After keeping its distance in 2004 and 2005, the Carnegie Mellon administration em-braced the Urban Challenge, sending out press releases announcing the GM partnership and promoting the involvement of some of its best researchers. CMU was America's robotics powerhouse, and it wasn't about to suffer another defeat. That, though, also meant a change. This was no longer just a Red Whittaker project. It involved various depart-ments, senior researchers, even a designated program manager to set up the GM partnership and track budgets and timelines. Carnegie Mellon was all in, and so the "Red Team" moniker would have to go. This outfit would take a name with the Scottish theme that graced the school's sports teams: Tartan Racing.

As the official team leader, Whittaker led the fundraising and higher-level duties, what tweed-wrapped academics called applying for grants and securing funding, and Whittaker referred to as "hunting, eating, and killing." He had always pushed his students to lead themselves. And leading Tartan day-to-day technical development would be one of Whittaker's protégés.

With his PhD in hand, Chris Urmson had left Carnegie Mellon for defense contractor SAIC, doing assorted robotics work like investigating new sensors. When Tartan formed up, CMU asked him to come on and offered him a faculty position. He happily accepted, and headed back to the old steel mill. The site had been renovated and turned into something closer to a proper testing facility, with a roof that didn't leak quite so much and the name "Robot City." There, Urmson set about teaching a robot to drive like a human.

The first step in what was now a familiar process was turning a regular vehicle into something that would respond to electronic pulses instead of hands and feet. To replace the Humvees that had become Sandstorm and H1ghlander, GM sent a pair of 2007 Chevy Tahoe SUVs to Pittsburgh, one black, one tan. Working with the Detroit engineers, the Tartan team tapped into the cars' engine control units and installed drive-by-wire systems that gave the computer control of the steering, brakes, and throttle. Jim Nickolaou, a General Motors engineer who pitched in on the effort, programmed the SUVs to stay in first and second gear, prioritizing acceleration over fuel economy.

The crew pulled out the third row seats to make room for racks of electronics, welded in roll cages in case they kept up the tradition of flipping robots, and installed heavy-duty brakes and run-flat tires. They put power outlets in the center console, so team members could keep their laptops charged during long stretches of testing. They festooned each SUV with an array of cameras, radars, and Lidar laser scanners, plus the klaxon siren that would sound whenever the robot was in motion. Both vehicles would be used for testing and development, but it was the tan Tahoe that would ultimately compete in the Urban Challenge. The team called it Boss, in honor of Charles "Boss" Kettering, the famed GM researcher whose hundreds of inventions included the all-electric car starter that made the hand crank defunct, an "aerial torpedo" (also known as a missile), and an incubator for prematurely born infants.

Their robot in working order, Chris Urmson and his team formed into smaller groups and started breaking down the Challenge. Reading through the *California Driver Handbook*, they realized that just because some driving tasks are easy for humans, that doesn't mean a robot can handle them. The Urban Challenge wouldn't include traffic lights, but it had something that revealed itself to be much scarier.

The rules governing a four-way stop seem simple: "Yield to the car which arrives first," the handbook said, "or to the car on your right if it reaches the intersection at the same time as you do." But in that sentence, the team recognized a problem. Say two cars arrive at the same time, from opposite directions. Who's on the right? What happens if four cars show up simultaneously, and each is to the right of another? Human drivers get by with hands to wave, horns to honk, headlights to flash. They express their intentions by inching forward or staying still. In the absence of hard-and-fast rules (and often in defiance of them), they negotiate, and mostly everything works out. Robots, however, aren't so versatile. The Tartan team found itself facing a new spin on Moravec's Paradox, struggling to get a machine to do something that for humans is so easy that it's hard to imagine it might be difficult.

Other problems were less existential but just as tricky. Merging into traffic required not only detecting other vehicles coming from various directions, but following them through space and time. In its simplest iteration, that involved comparing what the sensors saw in one moment to what they saw in the next, like a "spot the difference" game. Since the robot carrying those sensors was moving, it also meant precisely accounting for the change in perspective to properly compare objects, and dealing with the fact that those sensors didn't see everything, or see in much detail.

Navigation, too, looked to be a headache. The desert Challenges had come with a series of coordinates that created a route from start to finish. The Urban Challenge would work more like a scavenger hunt, giving the teams a map of the mock city and a series of destinations, leaving them to find the quickest viable path from one to the next. This work went to Dave Ferguson, who'd come to CMU after an advisor in his native New Zealand told him there was nowhere better in the world to build robots. Finishing up his PhD just as the competition kicked off, Ferguson was an expert in teaching machines to move through space, whether it was the "Groundhog" that mapped abandoned mines, a rover on Mars, or a robotic arm on a platform. Ferguson had worked under Sebastian Thrun before the latter went to Stanford, and had helped create an algorithm that let a robot quickly adjust its planned route as conditions around it shifted. It would be useful here: DARPA had said it would change the course as the Challenge went on, making prebuilt maps unreliable.

As the team assailed these and myriad other problems, they rebuilt the programs that had led H1ghlander and Sandstorm through the Mojave. They rewrote the entire framework, which dictates how all the bits communicate and makes sure everything runs properly, and added new skills the desert Challenges hadn't required. The vital task of managing how all this code came together fell to a newcomer named Bryan Salesky. Unlike nearly all of his teammates, he had left school after getting his bachelor's degree, which he'd earned at CMU's crosstown rival, the University of Pittsburgh. Heavyset, with a crop of red hair, Salesky was never much interested in research. He wanted to see his software having an impact on people's lives. After college, he went to Union Switch and Signal, developing software that kept trains from colliding in "dark territory," stretches of track not controlled by signal systems.

From there he went to the National Robotics Engineering Center, an arm of CMU's Robotics Institute focused on creating robots for commercial uses. When Tartan Racing came together, the CMU "elder statesmen" pulled him aboard as software lead.

Salesky was an easy choice for the role. He specialized in making reliable systems—a discrete skill from writing research-oriented code to prove an idea, not keep trains on the rails. He shared Thrun's worries about the "too-many-ideas syndrome." He was direct and forceful, good qualities on a team with the egos and politics that are a natural result of bringing academics together. He knew how to get people who liked working in the theoretical realm to focus on the practical questions at hand. And he brought a new skill set to a team that had previously been small enough that just Chris Urmson and Kevin Peterson could write most of the code, and keep track of it in their heads. With more than twenty people writing complex, integrated software, Salesky set up protocols for keeping everyone's work aligned.

He also proved a forceful lieutenant for Urmson. The two fought constantly: over technical approaches, over hiring, over strategy. Urmson came from the research world, and while he was one of the best people Salesky had ever met when it came to turning theory into science, he didn't have Salesky's training in turning science into a product. But they respected each other's points of view, and became good friends. A team racing to make one of the most skilled robots ever needed Urmson's ideas and Salesky's rigor. Some teammates took to calling them "Papa Bear" and "Mama Bear."

By January of 2007, with ten months to go before the Urban Challenge, Tartan Racing was well on its way. Its twin robots were running slalom courses amid the rubble and dirt mounds of the steel mill site, with the Pittsburgh skyline visible in the distance. They had completed multi-checkpoint missions and begun running through those devilish four-way stops. With the harsh winter setting in, it was time to head west. Not to the Nevada Automotive Test Center this time, but to the Phoenix suburb of Mesa, Arizona. Tartan's partner, General Motors, had invited them to its hot weather proving ground. At the expansive, well-equipped facility, the team took over an available workshop and fell into a familiar, brutal work schedule: wake up, work, eat whatever food someone hauled in, sleep, repeat. Six weeks before the start of the qualifying round, they decamped again. Back in 2003, racer Chip Ganassi had taught the Red Team to practice the way you race. Urmson figured DARPA was likely to host the Urban Challenge on a military base, since it would never be able to close down public streets for a robot melee, and he thought it would be nice to get away from GM's overly bureaucratic bailiwick. A quick internet search for disused bases led him to California's Central Valley.

A two-hour drive southeast of San Francisco, Castle Air Force Base had opened in 1941 as a flight school for American pilots headed into battle over Europe and the Pacific. When it closed in 1995, some of its nearly three thousand acres went to an air museum. Another chunk hosted a high-security prison. The rest lay mostly fallow, the soil contaminated by the fuels, oils, and chemicals that come with keeping warbirds in the air. But it had a large street network, a hangar to work out of, and room to run wild. So Urmson struck a deal with local officials, and the dozen or so core members of the team made camp in the abandoned officers' quarters. Conditions were grim. The place teemed with black widow spiders. One room contained a mess of feathers and bones a lunching coyote (or something) had left behind. When Mike Taylor, a Caterpillar engineer embedded with the team, noticed a pain in his foot one day, he took off his boot to find a pair of puncture marks in his pinky toe. His best guess was that a bat must have bit him in his sleep. On the advice of his father, a veterinarian, he called up the California Natural Resources Agency and asked if they had reports of rabid bats in the region. The answer came back in a worryingly grim tone: "Rabid bats are *everywhere*." Taylor headed to a doctor, took a rabies shot to the butt cheek, and went back to work.

The occasional wild animal interaction, in any case, wasn't as scary as the ordeals Bob Bittner dreamed up for Boss. An army and navy veteran who'd once worked as a foreman at the Pittsburgh steel mill the team had turned into a test site, Bittner didn't write code, but he'd worked with enough computing systems to know how to look for their weaknesses. As the Tartan testing lead, he created a twenty-five-page playbook that mimicked everything the govvies might throw at Boss: four-way stops where another vehicle went out of order, parking lots barely wide enough for the SUV to turn around, labyrinths that tested the bot's ability to navigate. To run these scenarios, team members would drive their rental cars or the handful of vehicles GM had sent over, rearranging the cones and barrels that represented assorted obstacles and the overturned plastic gutters that stood in for curbs. For the riskier tests, they used an inflatable car (nicknamed Stanley) on rollers. Bittner ended each session with a half-hour "free for all," where everyone did whatever he could think of to make Boss break a rule or otherwise trip up. Here, they were less worried about what DARPA concocted than what other teams' imperfect robots might do. They'd drive straight at the Tahoe, go the wrong direction down a one-way road, pull up alongside it and juke into its lane, and challenge it with whatever other nonsense they could imagine.

Those tests turned up software bugs and key information about where Boss was vulnerable. They also revealed the unexpected results that changes to a big, interconnected system can have, like an algorithm focused on intersection precedence that somehow deleted Boss's ability to execute three-point turns. In the final weeks before the race, the team turned its attention to hardening its system, and tracking down errors in the voluminous code. Sitting at the desks they'd set up in an old hangar, they celebrated each squashing of a software bug by hitting a red plastic button (akin to the Staples "That Was Easy" button) that played the Imperial March, the *Star Wars* tune that accompanied the approach of Darth Vader. Every repeat of the leitmotif signified another step toward the victory that had eluded them in the desert.

All those lines of code, though, wouldn't have been much good if it weren't for the hunk of aluminum riding on Boss's roof, spinning around ten times a second. After the 2005 Grand Challenge, Dave Hall had kept refining his sixty-four-beam Lidar sensor, ditching the lazy Susan–style design for one about the size and shape of a child's beach pail. Rather than competing in the Urban Challenge, he treated his former competitors as potential customers—some of whom now had a million dollars of DARPA's money to spend. Chris Urmson, Sebastian Thrun, and others had marveled at the Velodyne Lidar's 360-degree field of view, and resolution good enough to distinguish one team's robot from another's. It wasn't necessary for the desert crossing, but in the cluttered city, its ability to see the entire world in fine detail seemed likely to prove crucial. And for $75,000, anyone could have a HDL-64E to call his own. But it wasn't Dave Hall who helped install the Lidar. It was another familiar face from the Grand Challenges.

When he saw the data coming off Dave Hall's Lidar sensor, spinning away in the garage bay opposite his own at the 2005 Grand Challenge, Anthony Levandowski had immediately recognized its value. After the race, he got Hall to hire him as a salesman. The two men got along well, and made a good match. They were both in the San Francisco Bay Area, passionate about new technologies, and unabashed capitalists. Hall preferred his workshop to schmoozing with customers, and Levandowski, once converted to the gospel of Lidar, was a natural evangelist. They hooked what quickly became known simply as "a Velodyne" to the roof of Hall's pickup truck, and Levandowski took it around the country, showing it to all the Urban Challenge teams he'd met when he was just the crazy kid with the robo-motorcycle. Levandowski, though, was never much for full-time jobs, or putting all his eggs in one basket. While working for Hall, he joined another local outfit teaching robots to drive: Stanford's Urban Challenge team.

After winning the Grand Challenge, Thrun's gang had celebrated with a reception on campus and a photo shoot for *WIRED*. At Stanford's next big football game, Stanley drove the game ball onto the field. Its creators gave talks about how they'd done it and shared their work in a series of research papers. When the Urban Challenge rolled around, they jumped right back in, renewing their vows with Volkswagen and switching from Stanley the Touareg SUV to a hatchback Passat they named Junior. (The school's official name is Leland Stanford Junior University.)

Since desert testing no longer made sense and Palo Alto was short on abandoned steel mills, the team found its own disused military outpost in the naval air station at Alameda, across the bay from San Francisco. When they didn't want to face the traffic on the 101 Highway, they headed to the vast parking lot of the nearby Shoreline Amphitheatre. Along with the standard cones and barrels, Thrun's crew bought a cheap version of the machine that lays down street lane markers, to create a mock road network. Some days, they'd have to start off by cleaning up the broken glass and trash left by tailgating concertgoers. But they had all the room they needed. They'd pop on the white helmets they'd kept from their Grand Challenge testing and drive cars around Junior, honing its ability to track moving vehicles and safely move through intersections. For this race, though, Mike Montemerlo led the team on a day-to-day basis. Thrun was spending less time at Stanford, and a lot more in the glass-walled building across the street from their testing lot, known as the Googleplex.

Like Red Whittaker, Thrun had come out of the Grand Challenge eager to put his tech to work. He'd been in Silicon Valley for a few years, and had an itch to try his hand at being an entrepreneur. His mind kept returning to the camera footage Stanley gathered as it drove through the desert, which the team used as a reference whenever the robot stumbled. He liked to click through the data, transporting himself to various bits of the Mojave. In the summer of 2006, he asked one of his students to write a software program that could take such images and stitch them together. Impressed with the result, he recruited a small team and launched his first company, VuTool.

As it happened, Google cofounder Larry Page had been funding a similar effort since 2004, which would eventually be called Street View. Page had reached out to Thrun after attending the 2005 Grand Challenge, asking if the Stanford professor would like to see the robot he was designing in his free time. They went out for sushi, and in the parking lot of the restaurant, Page pulled out a remote control car he had been customizing, and asked for help with the navigation system. Thrun took the bot home, called his team in, and twenty-four hours later returned it to Page with a brand-new ability to chart its way around the world. It wouldn't be the last time he impressed Page with speed and skill.

Google's mapping team was using a van clad in $250,000 worth of sensors and unreliable hardware. They had added an immersive layer to its maps of Mountain View and Palo Alto, letting the user digitally "walk" down the street. Page saw mapping as a way to extend Google's reach from the digital to the physical world, and it was clear Vutool offered a cheaper, faster way to scale that expansion. In March 2007, Thrun was talking to various Silicon Valley venture capital firms, unsure of whose money to take. Page offered to bring him into the Google empire. Thrun hesitated—he liked the idea of running his own company—but one morning, during a student presentation, he checked his email to find an offer he couldn't turn down. It came with whopping signing bonuses for him and the Vutool team, with the promise of more, depending on how fast their tech could grow. And Page had an appetite: He asked the team to map a million miles.

For help, Thrun hired Levandowski, who was starting to display a Forrest Gumpian knack for being in the thick of the autonomous vehicle action. Thrun was impressed with the thinking behind the Ghostrider motorcycle (though he gleefully showed video of it diving into the pond in talks he gave). Levandowski, Thrun saw, had a remarkable gift for getting things done. As a no-name independent entry, he had built a team, raised money, attracted sponsors, and produced a robot nobody else had had the temerity to try.

Charged with scaling up the mapping operation, Levandowski went out and bought one hundred cars. Thrun caught some flak for ignoring Google's expense report protocols, but he didn't mind—his project was charging along. To get the cars, each with about $15,000 worth of equipment, into the street, Levandowski hired drivers off Craigslist. In just seven months, they hit Page's million-mile milestone.

Just as DARPA had finished with the desert, it had finished with the California Speedway, which had hosted the qualification rounds for the first two Challenges. When the thirty-five teams invited to try for the final round of the Urban Challenge gathered in the last week of October 2007, they found themselves in the latest strange world of the Pentagon's creation.

DARPA program manager Norm Whitaker had known that running the qualification or final round anywhere near civilians was out of the question. The success of the 2005 Grand Challenge didn't erase from anyone's memory the fact that these multiton hunks of metal were capricious beings, prone to spastic swerves and uncommanded starts. After being turned away by various active military bases, Whitaker landed on George Air Force Base in Victorville, California. A two-hour drive northeast of Los Angeles, the airfield was built during the run-up to World War II, then shuttered as the Cold War fizzled out. Its squat, yellowy brown stucco houses were packed with asbestos and pocked with bullet holes courtesy of the soldiers who'd come here for urban warfare training. It appealed to DARPA for the same reasons the Tartan team had moved to the Castle base. It had a street network, plenty of space, and nary a desert tortoise to worry about trampling.

To populate this dusty Potemkin village, Whitaker hired a crew of Hollywood stunt drivers and folks from SCORE, Sal Fish's off-road racing organization. They would create the traffic the robots had to navigate, using a fleet of fifty Ford Taurus sedans retrofitted with crash-resistant roll cages. To get them used to driving around autonomous vehicles, Whitaker brought them to a movie theater parking lot near

DARPA's Virginia headquarters. He designated one car as a "robot." Its driver sat blindfolded, following the directions of a colleague sitting shotgun. Whitaker figured the resulting herky-jerky driving mimicked an unpredictable machine, as the other drivers maneuvered around it.

The qualification round began the morning of October 26, 2007, after an opening ceremony that included the national anthem and the introduction of the thirty-five team leaders. Over five days, the robots cycled through three courses, labeled A, B, and C, designed to test for different capabilities. A was all about making left turns into traffic provided by those stunt drivers. If a robot was too aggressive or too timid, it would be penalized. B resembled the qualification course from the Grand Challenges, a winding path on which a robot would have to dodge cones, barrels, junked cars, and a mock construction zone. It also held the parking lot, where the bot would have to pull into an open spot, then back out again. C was known as the "belt buckle" for its layout. One road made a square, with another road running straight through it, forming two intersections. This was where the machines would face the horribly human four-way stop.

To suit the difficulty of the task, Tony Tether had pomped up the Urban Challenge. The grandstands had room for fifteen hundred spectators. A twenty-one-thousand-square-foot tent could fit eighteen hundred, plus a stage. Those who couldn't make it to Victorville could watch the race via webcast, with *MythBusters'* Jamie Hyneman and Grant Imahara as commentators. The one hundred officials staffing the event worked out of trailers. Without counting the prize money, DARPA had spent more than $21 million putting the race together, twice what the 2005 Grand Challenge had cost.

Over five days of qualifying, each robot went through the various parts of the course multiple times. For the teams at the event, most of the action was hard to see. In between snacks provided by Chip Ganassi, the racer who had become a Red Whittaker devotee, the Tartan Racing crew tried to track Boss's location on the course by listening for its distinctive siren. As the qualifications moved along, hearing the sign of life that confirmed that the robot was still running became increasingly comforting.

It was clear not every team was up to the Challenge. On its first attempt, Team Gray, the Cinderella story from the 2005 race, failed to get past the first curve on the road course. Team Jefferson's Toyota Scion lost its windshield to a railroad gate. One robot cleaved the Velodyne Lidar off its roof when it drove under a carport while attempting a U-turn. Team Golem's Toyota Prius smashed into a concrete barrier at 30 mph.

The Oshkosh military truck—this one a smaller version with four wheels instead of six—hit and dragged a car several yards through the faux parking lot. It wasn't disqualified, giving weight to the complaints of the smaller, independent teams that Tether was playing favorites. The DARPA chief gave his own reason for limiting the field to the more experienced efforts. At one point during the week, he and a few DARPA officials were driving on the course when the robot they were following swung around and barreled straight for them. As they were all yelling "Pause!" either the kill switch kicked in or the bot detected the upcoming, human-filled obstacle. It stopped with a few feet to spare, reminding Tether of what the stakes really were.

Among the disqualified was the tan Jeep known as the Spirit of Kosrae. When the Urban Challenge rolled around, Melanie Dumas and her Axion Racing teammates signed up, but the fervor that had propelled them into the Mojave had dissipated. They did less testing, spent fewer late nights writing code. Dumas's software skills landed them a spot in the qualifying round, but it was clear they were outmatched by the teams stocked with full-time researchers and funded by Fortune 500 sponsors. Trying to turn left into traffic, their Jeep failed to spot one of the human-driven cars and hit it. The driver was fine and got a steak dinner for his trouble, but Axion's run was finished.

Tony Tether had always liked the team. Axion's ragtag band was exactly what he'd had in mind when he opened the Grand Challenge to anybody. He was especially impressed by Dumas, and when he took her aside to deliver the news of her team's demise, he followed it with some unexpected advice. She should apply for a job with DARPA, he said. The Pentagon's research wing could use people like her. A few months later, Dumas became the latest in a long line of DARPA program managers charged with sniffing out the next great technological breakthrough.

During its time on the course, Carnegie Mellon's Boss had no major problems, though it did at one point pop into reverse and accelerate for a good thirty meters—the result of a bug the testing at Castle hadn't turned up. Stanford's Junior proved overly cautious, waiting for so long to turn left into traffic that Tether told Mike Montemerlo he could either code in some gumption or pack his bags. Both robots, though, were good enough to qualify for the final round. So were five other of the eleven "Track A" teams. Perhaps more impressively, four "Track B" teams—the underdogs who hadn't received a million bucks from DARPA—also qualified.

By this point, the tech universe was obsessed with DARPA's autonomous Challenges, dreaming of the day when the futuristic vehicles merged into their everyday lives. The competitors were profiled in *Time* and *Forbes*. They appeared on ABC. *Popular Science* had reporters on the ground, writing live dispatches. For the main event on November 3, Google founders Larry Page and Sergey Brin showed up with a private plane full of executives. They were there to cheer on the various teams they had sponsored, chief among them the Stanford crew staffed with its own Street View team, including Sebastian Thrun and Anthony Levandowski.

To win the Urban Challenge, the vehicles would have to complete three mock supply missions in the simulated city, one in each of the three courses. After each, the bot would come back to "base," where its team would load in the next set of coordinates and goals. This being the final, DARPA would add more traffic and make the parking lots bigger and harder to navigate. Around the "belt buckle," with the four-way stops, DARPA would block one route with an old car, forcing the robot to recognize the obstacle, make a U-turn, and find another way—then ace the intersection.

For the third consecutive Challenge, Red Whittaker's robot had taken pole position. The team dressed Boss in black livery and put GM's blocky blue logo on the hood, firmly tying the corporation to the effort. The SUV wore nineteen sensors, eighteen of them playing second fiddle to Dave Hall's Lidar. The robot could calculate more than one thousand possible trajectories every second. It was ready to take its revenge after the failure of the first two Challenges. But just before 8 a.m., moments before Tony Tether dropped the green flag, the entire Tartan team nearly had a heart attack. Boss's GPS signal, vital to its navigational skills, was gone.

"Not again," Chris Urmson thought. Another mishap at just the wrong time. The crew tasked with starting the robot scrambled. They swapped out the GPS's various parts. They checked over every system, crawling all over the Tahoe, desperate to find whatever had changed since they had run so cleanly in the qualifiers. Most of the team sat in the grandstands, watching the trouble but unable to help. The computer's boards weren't staying in place, they heard. Or the emergency stop system was busted. Larry Burns, the exec who had gotten General Motors involved with the team, looked at the GM logo on the frozen robot's hood, sweating in the cool California air.

After a few minutes, the team landed on the culprit. For the pleasure of the fans, DARPA had set up a Jumbotron by the starting gate, to display scenes from the event. Boss's pole position put it right next to the thing, close enough for the screen's radio signal to scramble the Tahoe's GPS. Tether ordered someone to kill the giant screen. The GPS sprang back to life, and the Carnegie Mellon team went to restart its robot. It would take about half an hour for Boss's systems to come fully online, so Tether decided the show would go on. The first robot to enter the "city" was Virginia Tech's Odin. Then went Stanford's Junior. Boss ended up leaving the gate after nine other vehicles. As with the desert

Challenges, each vehicle would be judged by how long it took to complete the course, not by the order in which it finished.

Once more, DARPA gave each team a set of waypoints to follow
(by 2007, the agency had abandoned CDs for USB thumb drives).
And once more, some of the robots got themselves into trouble. An
old Subaru entered by the University of Central Florida smacked into
a house. Oshkosh's military truck was a few feet from hitting a pillar
before a DARPA official hit its kill switch. Four cars reached an intersection at the same time, leaving each in confusion about whose turn it
was to go, creating history's first robot traffic jam. MIT's car tried to pass
Cornell's stopped SUV, only to cut the move too tight just as Cornell
started moving again, yielding what was likely the world's first autonomous vehicle fender bender.

When it finally got its chance, Boss got out of the gate in a hurry,
nailing a quick left turn and then a right to get onto a winding path.
Looking like a driver in a rush, but one who doesn't want to break any
rules, it jerked to a halt at each stop line, then, once it knew it could
move ahead, hit the accelerator, hard. Whittaker wanted it that way. He
was about as impressed with the idea of a race with speed limits as a
sprinter watching Olympic speed walking. Boss would stay below the
limit, but there weren't any rules against aggressive acceleration and
braking—and there was no human inside to get queasy.

Stanford's Junior moved at a more reasonable pace. Pulling up at a
T-intersection, it waited patiently to head left, its turn signal blinking
and its Velodyne sensor whirring as it watched one car after another
zoom past. When the road finally cleared, it accelerated, executing a
smooth left turn and moving along. When it entered the large roundabout that connected various parts of the course, though, it went past
the turn it should have taken, and kept circling. Its makers' minds leapt
to the worst case scenario, that some bug had scrambled Junior's ability to handle this situation, that it would drive in circles until DARPA
put the dizzied robot out of its misery. The truth was stranger: After
one go-round, Junior executed its turn. Every time it went through the

roundabout that day, it did the same thing, making one full circle before heading to where it should go. Yet more proof that robotics was not just difficult, but often baffling.

Over the next six hours, the eleven vehicles tackled one task after another, each following its own order through the missions. Odin, the Ford Escape built to drive itself by Virginia Tech's Team Victor Tango, cruised into the parking lot area of the course, found its spot, and pulled in calmly. After a beat, it backed out again, then returned to the main bit of the course for its next task. Boss turned left into traffic. Junior passed a stopped car in its lane, then shifted back to the right. The fifteen hundred spectators in the grandstands and in the tents were watching a race that looked nothing like NASCAR or Formula One. The "race" looked a lot more like people driving around a small suburb. It could be awfully boring—until those watching remembered that they were in fact witnessing a bunch of robots navigate the trivial but complex rules, rituals, and customs of human driving. All by themselves.

At 2 p.m., Stanford's Junior became the first robot to cross the finish line, having completed its three missions. Boss rolled in a few minutes later, followed by Virginia Tech's Odin. But finishing order was not destiny. The robots would be graded on how long they had spent on the course, and how well they had driven. The next morning, the teams gathered in front of a stage, where Tether stood with three giant checks wrapped in brown paper. After brief remarks, he revealed third place:

Virginia Tech's Odin.

Then second:

Stanford's Junior.

And finally, first place in the 2007 DARPA Urban Challenge, the vehicle that had averaged one mile per hour faster than Stanford's, the one programmed to leap off the starting line and screech to every halt, that didn't waste time taking extra loops through the roundabout:

Carnegie Mellon's Boss.

As the crowd clapped and hollered, Red Whittaker ran onto the stage to claim, at last, his victory. After four years of brutal work and dashed dreams, he looked dazed. "That's the first time I've seen Red speechless," Tether said. But Whittaker never stayed down for long. "Give me the mic," he shot back.

Standing on stage, holding his huge bronze eagle trophy, Whittaker was far from ready to end his career. But he was ready to pass the autonomous driving torch to the likes of Chris Urmson and the students he had turned into an army willing to take on the impossible.

"I envy the timing of people coming into robotics now and reflect in many ways I was a little too early for the game," he later said. "I am from the era of the great imposters, meaning that it was before the time of robotics credentials in the same way that the Wright brothers would be before the time of aeronautics degrees. . . . The opportunity now is to engage and immerse so wholly, and to do so from youth. That's what makes the great ones."

From the stage, he looked at his apprentices and colleagues, squinting and sweating under the California sun. He had played an instrumental role forming them into a team. It would be their job to bring what they had together created into the real world. After glad-handing a small horde of dignitaries and sponsors, Whittaker climbed into his rental car and headed for Las Vegas to catch the red-eye back to Pittsburgh. When he got to the airport, still wearing his Tartan Racing gear, someone asked him, "Wow, were you in that car race thing?"

"I won that race," he said. Then he flew home and went back to work. With a long history in planetary robotics, Whittaker had been itching to make a lunar rover. Google had recently announced it was funding its own version of the DARPA Challenges, the Lunar X Prize. Having conquered the desert and the city, Whittaker wanted to send a robot to the Moon.

"I always thought autonomous vehicles would happen," Axion Racing's Melanie Dumas said. "To be honest, I didn't think it would happen in my lifetime." She was barely thirty years old when the cars that braved the Urban Challenge offered proof that this long imagined technical marvel was on its way. That research done over decades was ready to bear fruit. That cameras and radars and a new spin on lasers could reveal to machines a complex world, that software and ever more powerful computers could lead them through it. That passionate engineers could attack the deep-seated difficulty of robotics and come away victorious, even if not all of them got to wave aloft a giant check. That DARPA, an institution funded by the people, could still deliver startling innovations. That with the right teachers, collections of metal and plastic and silicon could learn to drive themselves.

The revelation didn't reach far. The press coverage faded quickly and the public's attention swung to the bigger questions of late 2007. The George W. Bush era was drawing to a close and voters were looking for a new leader. American men and women continued to fight in the Middle East. Sharp-eyed observers saw trouble coming for the economy. But like the witnesses at Los Alamos who watched the first test of the atomic bomb melt the desert sand into glass, those gathered at the dusty, decommissioned air force base that November day knew they had seen something awesome. When it would change the world, and how, was up to them.

The Continuous Representation of Reality

WHEN KYLE MACHULIS DROVE INTO THE PARKING LOT OF THE San Jose DoubleTree in January 2009, the people lined up for the fursuit parade represented just one-fifth of those attending the Further Confusion conference. A furry, to use a simple and broad description, is a fan of anthropomorphic animals. Most furries create characters, usually cartoonish animals with human characteristics, whom they inhabit online or in real life. The San Jose gathering was among the largest of its kind, largely because furries formed a surprisingly large percentage of Silicon Valley's tech workers. Getting online was the main way for a furry to connect with her fellows, and when the subculture arose in the late 1980s, that meant knowing how to run servers and build websites. When the dot-com bubble seized Silicon Valley in the mid-1990s, the skills furries had learned to connect with each other became valuable to companies with chairs to fill. The economic frontier of Silicon Valley embraced unconventional lifestyles. Nonconformist ways of living often come with nonconformist ways of thinking, and nothing gets upended—or disrupted, to use the local lingo—by people who think like everyone else. More to the point, a worker's skill with a keyboard mattered more than however they spent their free time. For all those skills, though, few of the furries in San Jose that day knew what they were posing for as Machulis drove around them in a Nissan pickup whose bed carried what looked vaguely like tornado-tracking equipment. Lidar was still a little-known technology.

A few feet above the truck bed, mounted on a metal trestle, the coffee can–shaped Velodyne Lidar twirled, bouncing pulses from its sixty-four lasers off whatever was nearby. Simultaneously, a GPS tracked the truck's position, an inertial measurement unit noted its direction, and wheel encoders reported the vehicle's exact speed. They all piped their results into a bright yellow box, about a foot cubed and sitting next to the Velodyne, which correlated each bit of data to the millisecond at which it was produced. Back in his office, Machulis transformed the returns into a digital, three-dimensional recreation of whatever he had seen. It was detailed enough to pick out the pleats in Minnie Mouse's skirt.

Machulis had borrowed the truck from his employer, a mapping and robotics company named 510 Systems. Based in a brick-facade, two-story building in downtown Berkeley (and named for the city's area code), 510 had a lot in common with the other tech startups of the time. It was staffed mostly by young men, had no concept of an HR department, and took a liberal attitude toward how employees should file expenses. Staffers who were into rock climbing scaled the rafters; others brewed beer in the office. They took turns riding through the office on a "haunted" Segway with a tendency to glitch and buck its rider. But 510 had a few unusual traits. It was across the Bay from San Francisco, an hour's drive from Palo Alto and Mountain View, the Silicon Valley cities where many new companies set up shop. It dealt with complex hardware, when many were getting rich by developing mobile games like FarmVille and Candy Crush. It divulged little of its workings to outsiders. It included so little information on its website that one job candidate thought 510 must do some sort of secretive defense work. And its CEO was a woman in her mid-fifties named Suzanna Musick, whose background was in marketing. But she didn't set the company's direction or priorities. The real boss was her stepson, who had moved in with her and his father as a teenager, after spending his childhood in Belgium.

Anthony Levandowski founded 510 Systems with fellow Berkeley grads Pierre-Yves Droz and Andrew Schultz on May 10, 2007. Levandowski at the time was working for Sebastian Thrun on Google Street View, helping on Stanford's Urban Challenge team, and selling Velodyne Lidars for Dave Hall. After the Urban Challenge, he took a job studying unmanned technology with Ensco, the defense and aerospace contractor whose bathtub-like robot had flipped over sixty-six seconds into the 2004 Grand Challenge. He soon tired of the work, uninterested in writing reports that, if successful, would land him a contract to write more reports. Though he kept his name off much of the 510 paperwork, this new company was his creation. And it was making technology that promised to change how its customers saw the world.

Millions of Lidar returns, camera images, and GPS coordinates aren't much good for mapping unless you can link them together, the way the human brain automatically links what its body sees, smells, hears, tastes, and touches into a cohesive understanding of its surroundings. That's what 510 Systems did. It developed the IP-S2, or Integrated Positioning System. This was the yellow box amid the sensors on the truck Machulis took to the DoubleTree in January 2009. It collected and correlated the data from those sensors. 510 wrote the software that took the results and turned them into a 3D visualization a person could inspect and analyze. The result was a powerful new tool for creating a high-fidelity record of a physical space, to be used by anyone who needed to know where things were and how they moved over time.

With that kind of intel, utility companies could see where tree branches threatened their power lines. Transportation departments could keep track of road signals; construction crews could measure their progress at work sites. Conventional surveying techniques involved workers setting up tripods in carefully selected locations and collecting data point by point. With the IP-S2, anyone could instead mount the surveying rig on a vehicle, drive through the area of interest, and click through the 3D map that 510's software spat out.

To put the tool in customers' hands, Levandowski paired with Topcon, a Japanese company founded in 1932 to make surveying instruments, binoculars, and cameras for the Imperial Army. After World War II, Topcon went civilian, eventually expanding into California and making a variety of high-tech tools for the medical, construction, and agricultural industries. Levandowski had first reached out to the company in his search for sponsors before the 2004 DARPA Grand Challenge. Topcon exec Eduardo Falcon was charmed and gave Levandowski his first cash donation, of $20,000, along with a GPS. No matter how well the Blue Team did, Falcon figured, they would stand out for having the only motorcycle. When Levandowski came back a few years later with this venture, Falcon was happy to reconnect.

For the IP-S2, 510 Systems worked as a sort of external R&D lab, figuring out how to make the thing work and customizing it for different uses. Topcon would manufacture and sell the result, promising customers what it dubbed the "continuous representation of reality"— a detailed, accurate, and near real-time portrayal of whatever it was they wanted to see. Customers for the "Topcon box" included Microsoft, which was working on its own location imagery technology. Nokia-owned mapping company Navteq used the box to develop a turn-by-turn navigation offering. One of the biggest buyers was Google, which used the IP-S2 for Street View. Which was odd, because Levandowski hadn't left Google when he started 510. He was still working on the hardware side of Street View, and now he was sending his employer's money across the Bay, into his own pockets. The Google bosses, including Sebastian Thrun, knew about this unorthodox setup, but at one of the world's most profitable companies, who cared about the cash? Levandowski, as hard-charging as ever, was getting the job done.

The workers at 510 knew Levandowski worked for Google, but didn't have much in the way of details. Campbell Kennedy, a 510 product manager, understood a bit more when the Street View team was having trouble with the IP-S2 and asked if someone from 510 could help them out in person. He went to Mountain View with Levandowski and Suzanna Musick. When the three walked into the room for the meeting, Levandowski sat down not with them, but with the Googlers—the Googlers who then grilled Kennedy about 510's technology.

Levandowski's Google commitment—including regular trips to Hyderabad, India, where two thousand workers turned the Street View cars' data into useable maps—kept him away from 510 for long stretches. When he was around, usually in the evening, he had the same effect on his employees he'd had on his DARPA Challenge teammates. He was charismatic, full of ambitious ideas, and unimpressed by practical hurdles. One day, 510 software engineer Ben Discoe was sitting in the office kitchen, musing over how great it could be to have a 3D visualization of his family's Hawaii tea farm. Levandowski wandered in, overheard, and told him to just go do it. A few weeks later, Discoe was flying west with tens of thousands of dollars' worth of 510 equipment. Another engineer, Willy Pell, found conversations with Levandowski almost addictive. The 510 founder understood their tech at the detailed level, but could also weave disparate ideas into big-picture imaginings. They'd have long brainstorming conversations, covering whiteboards with ideas. This was where Levandowski was at his most compelling: in small groups where he could connect with people one-on-one, waving his long arms and urging others to debate with him, may the best idea win.

It made for an exciting atmosphere but an inefficient business. The company engineers reported sporadic direction from the top, and were moved from project to project at a moment's notice. One post on the business-focused reviews site Glassdoor complained, "Not much information from management about the direction of the company." Another read, "Uncertainty and doubt as to what is going on. Management does not tell the truth to employees. Overall direction of the company is unknown." Pell compared 510's structure to a subway map's tangled, intersecting lines. They had hardware and software efforts, and a stable of customers with particular needs. One team built a version of the sensing rig that fit into the belly of a Cessna airplane, so Google Maps could include 3D images of cities taken from the air. On the building's second floor, 510 cofounder Pierre-Yves Droz started developing a new Lidar, to end the company's reliance on Dave Hall's $70,000 sensors. The company helped produce the music video for the Radiohead song "House of Cards," which was shot in Lidar. Its engineers helped Topcon develop controls systems for automated farm equipment. Yet another node on the subway map shared a building and some employees with 510, though it in fact belonged to another company, which Levandowski had incorporated on June 12, 2008. On the second floor of 510's Berkeley office, Anthony's Robots was building a car that could drive itself.

For Levandowski, the conclusion of the DARPA Urban Challenge in November 2007 was just a starting point. From his first days building his robotic motorcycle, he understood that autonomy could fundamentally change how everyday people moved through their lives. So when a producer from the Discovery Channel asked him if he could use his autonomous motorcycle to cart a pizza through San Francisco, Levandowski wanted in. He knew the producer from his 2007 appearance on the Animal Planet show *Chasing Nature*, in which student engineers mimicked the skills of various animals. In an episode focused on the jumping ability of dolphins, Levandowski traded his "Blue Team" T-shirt for a wetsuit, flippers, and a back-mounted air cannon. Towed behind a boat, he fired the cannon and flew twenty feet through the air before splashing down, face-first. No other contestant soared farther.

Now the producer was working on a Discovery Channel show called *Prototype This!* In each episode, four "inventors" took on an outlandish challenge by building outlandish machines. In their one, thirteen-episode season, they made a flying mechanical lifeguard, a wearable airbag for workers building high-rises, a water slide simulator, and a pair of huge boxing robots. They filmed the show on Treasure Island, a four-hundred-acre man-made landmass in the San Francisco Bay that the Army Corps of Engineers had built to host the 1939–40 Golden Gate International Exposition, with the idea to convert it later into an airport for the Pan Am Clipper seaplane. The navy took it over during World War II and stayed until 1997. In 2008, Treasure Island was a lonely place, home to some low-income housing and a legally acceptable amount of radiation deposited by ships exposed to nuclear testing during the Cold War. But it was the distance from San Francisco—you had to drive half the length of the oft-congested Bay Bridge to reach the island—that made it hard to get food delivered there. This genuine source of frustration for the *Prototype This!* crew was the premise for the show's eighth episode, "Automated Pizza Delivery." Over the forty-five-minute episode, the gang tried tossing pizzas out of a low-flying blimp and customized a rolling robot to carry the cheesy pies. Then they turned to "software engineer, autonomous vehicle expert, and all-around whiz kid Anthony Levandowski."

The show's narrator didn't mention that Levandowski played a role in Street View, one of Google's most audacious and beloved programs. Levandowski had brought the *Prototype This!* pitch to his bosses for approval. Google didn't want its name anywhere near testing unproven tech on public roads for some goofy show. He could go ahead, but was to keep his relationship with the tech giant out of it.

The producer's original idea was to use Levandowski's autonomous motorcycle, but Ghostrider wasn't available. Along with Stanford's DARPA Grand Challenge–winning Stanley and Dave Hall's original Lidar, it was now part of the Smithsonian's permanent collection, officially sanctioned as a piece of technological history. Levandowski used two Toyota Prius hybrids instead. The first was equipped with a suite of lasers and the 510 Systems–designed IP-S2. With Levandowski doing the driving, it mapped the delivery route from downtown San Francisco, onto and across the Bay Bridge, then down the ramp onto

Treasure Island. The second car was the star of the show. In a Berkeley driveway not far from the one where he and his Blue Team had worked on Ghostrider, Levandowski showed off Pribot, with a rooftop Velodyne Lidar and a trunkful of computers.

The *Prototype This!* narrator didn't reveal that just a few weeks before the shoot, Pribot hadn't existed. It was the product of a feverish push by Levandowski, his 510 cofounder Droz, and 510 product manager Campbell Kennedy. In a throwback to the DARPA Challenges, they went days without sleep, racing to make their robot work before the fateful day when it would be put to the test. They started by tapping into the Prius's CAN bus, effectively the car's nervous system, which would let them control it with electronic commands. That worked fine for the acceleration and steering, but reverse engineering the braking proved trickier. The Toyota software that ran the electronic stability control and antilock braking made the effort so complicated and time-consuming that they opted instead to install a hydraulic motor on the floor of the car to physically push the pedal. Without a proper proving ground, they tested their work on the roads of Berkeley at night, always with someone in the driver's seat, ready to hit the kill switch.

The hastily built robot, with a 510 Systems logo covering its hood and "Topcon" written across its bumper, could not read traffic lights or detect stop signs. It couldn't have handled the traffic that faced the bots in the Urban Challenge. Since it only had a rudimentary collision detection system, it might not have been able to complete the simpler Grand Challenge, either. It only stood a chance of making the journey through San Francisco because the Discovery Channel producers arranged to have the police shut down the streets that made up its route, including half the Bay Bridge.

Challenge day came on Sunday, September 7, 2008. A little before eight in the morning, Levandowski set up Pribot on the Embarcadero, the boulevard that follows San Francisco Bay's curving shoreline. The car hesitated as it pulled away from the curb, then put itself in the middle of its lane, its steering wheel twitching and its Velodyne Lidar spinning. It followed GPS waypoints, much like the DARPA bots that had competed in the Mojave. This robot, though, traveled in the middle of a police motorcade and right behind a flatbed truck, from which Levandowski looked back, ready to hit the remote kill switch. It wasn't the smoothest performance. Pribot strayed to either side of its lane at points and took wide turns. But it slowly made its way through the city and up the winding ramp that led to the lower, eastbound deck of the Bay Bridge.

Holding the kill switch in his right hand and high-fiving the show's hosts with his left, Levandowski watched his creation cross the span. But when the car reached the exit for Treasure Island, the cheers turned to groans. As Pribot curled down the hairpin ramp, it edged left, scraping against the concrete railing and pinning itself stuck. Levandowski shut it down, folded his lanky body through the driver's side window, and drove the car the rest of the way off the ramp. Once safely on Treasure Island and back in autonomous mode, Pribot made the final few turns into the *Prototype This!* office parking lot. Each hoisting a slice of freshly delivered pizza, Levandowski and the show's hosts declared a toast "to the first ever autonomous pizza delivery vehicle." More than that, it was, almost certainly, the first time an autonomous vehicle had explored public roads with nobody behind the wheel.

Soon, though, Levandowski would be part of a much bigger, more serious effort to deliver on the promise of a technology the DARPA Challenges had shown was possible—one that wouldn't end when the credits rolled.

Larry Page had been interested in autonomous vehicles long before he attended the DARPA Challenges. As a twenty-two-year-old student in Stanford's computer science doctoral program, he'd considered the technology as a research idea. Instead, he followed his advisor's advice and focused on search, finding a new way to understand the links between the disparate regions of the World Wide Web. A decade after founding Google with his classmate Sergey Brin, he was starting to push his company's influence beyond the limitations of screens and keyboards. Sebastian Thrun (aided by Anthony Levandowski) had proven an effective conquistador, accelerating the Street View project and then spearheading an effort called Ground Truth, in which Google built its own global database of maps from scratch. The massive project, almost unfathomable in its ambition and relying on the work of thousands of people, freed Google from the need to license the data of existing cartography companies and formed the basis of all its ensuing mapping efforts. It also gave Thrun what he called "an appetite for scale" that drew him further and further from his past life as an academic.

Now Page wanted to put computer-controlled cars on the streets his company had mapped. The influence of the *Prototype This!* episode on his thinking is hard to discern. Some insiders believe the creation of Pribot was a bid to push for a robotics project at Google, a public stunt by Levandowski and his boss, Thrun, to show Page they intended to tackle autonomous driving one way or another.

In Thrun's telling, it was the opposite: that Page was intent on pursuing the tech, and Thrun tried to wave him off. The DARPA Challenges were one thing, with their limited conditions. No traffic lights, no pedestrians, no unexpected construction crews, no crushable tortoises. Trying to build a robot that could navigate the real world, in all its complexity and chaos, was folly, Thrun said. Page persisted, asking for a technical reason why it was impossible. Frustrated, Thrun snapped. "It can't be done, goddammit," he said. Then he went home, tried to think of that specific, technical reason, and came up empty. When he admitted as much, Page talked up the economic view that Levandowski had grasped heading into the first Grand Challenge: Transportation formed an enormous chunk of the economy, and applying autonomy to even a sliver of it could make for a massive business. They could make a company as big as Google itself, Page said. Even if it had a one-in-ten chance of working, it was worth the investment of money and time. Whatever the exact motivation, Thrun agreed to give it a go, and he knew exactly who he wanted to join him. He had met them in the Mojave Desert.

In October of 2008, six men met Thrun at his Lake Tahoe home. All were veterans of the two teams that had dominated the three DARPA Challenges. From Thrun's Stanford effort, there was Mike Montemerlo, Hendrik Dahlkamp, Dirk Haehnel, and naturally, Levandowski. Chris Urmson and Bryan Salesky represented Carnegie Mellon. As rivals, they had pushed one another to create some of the world's most sophisticated autonomous vehicles. Now Thrun wanted them to come together as one team and finish what they had started outside the Slash X Cafe.

Google, though, seemed an odd place to do it. It was a search engine company. It did email and maps and owned YouTube, but it didn't make any hardware. (That changed in June 2012, when Google introduced a digital media player called the Nexus Q. Today it makes phones, tablets, laptops, and more.) Street View and Ground Truth charted the physical world, but they didn't interact with it the way an autonomous car would. This would be a totally new direction for the company. And how could the DARPA veterans be sure Page wouldn't lose faith in this project after a year and kill it? They'd be left without jobs, having abandoned promising, dependable careers for a risky Silicon Valley venture.

Thrun had counterarguments. Google might be about software, but it was no ordinary company. When it had gone public in 2004, Page wrote a letter warning potential investors that he and Brin had no intention of bowing to short-term financial objectives. "We may do things that we believe have a positive impact on the world, even if the near term financial returns are not obvious. . . . We will not hesitate to place major bets on promising new opportunities." Google was awash in cash—it made more than $4 billion in profits in 2008 alone—and could afford to fund the effort, even if no one knew what it would cost or how long it might take. And in case Page did give up on the idea, Thrun designed a compensation package with lavish (especially compared to the academic world) salaries and bonuses that would help them get through any stretch of unemployment. If they did their work well, the team's early members would make upwards of $1 million in a year.

Most of all, Google was offering them the chance to be pioneers. Even the roboticists among them hadn't imagined working on cars before Tony Tether started talking about a race in the Mojave. After the Urban Challenge, Chris Urmson and Red Whittaker had proposed to General Motors that they continue their autonomy partnership, building on what they had done in the mock city. The Detroit automaker declined. Even a year before the financial crisis that would knock it into bankruptcy, GM's business was imploding. In 2007, it lost $38.7 billion, an auto industry record. Perhaps the only company pursuing fully autonomous vehicles was Caterpillar, which had hired Urmson and Salesky to develop robot trucks for working in mines.

Thrun offered these engineers the opportunity to do something they and everybody else considered just a bit nutty. To take their DARPA Challenge work out of the desert and put it in the middle of the world where they lived. To give them a few stories to tell their grandkids.

Thrun's sales pitch wasn't enough to convince Salesky, who wanted to see the Caterpillar project through and wasn't ready to abandon Pittsburgh for Silicon Valley. Thrun did win over most of his Stanford teammates, who were already in California, most of them working for Google in some capacity, and Chris Urmson, who was game for a change of scenery, a new mission, and a promotion. Thrun would be the head of the projects, but when it came to the day-to-day work of teaching a car to drive, Urmson would be in charge, if unofficially. He resigned his role on the Caterpillar team, gave up the professorship he had just started at Carnegie Mellon, and headed west.

Google's vision was grand in its ambiguity. Page wasn't proposing a typical government or defense industry contract. He wasn't looking for a specific solution to a specific need, with carefully defined funding and spending limits. He just wanted an autonomous car, and he was willing to write something that looked a lot like a blank check to get it. He did, however, believe in aggressive goals. In negotiations with Thrun, he and Brin set two targets for the team, one to focus on scale, the other on skill. The first was to accumulate one hundred thousand miles of autonomous driving on public roads. Those, they could log wherever they liked. Then the cofounders pulled up Google Maps and selected ten routes, each roughly one hundred miles and all within California. Thrun and Urmson's team would have to make a car that could handle every inch of each route, without human intervention, about a thousand miles in all.

"They took great pleasure in picking really hard roads," Thrun said. Roads like San Francisco's famously double-jointed Lombard Street. Roads through downtown Los Angeles and the Sierra Nevada mountains. Roads tracing the winding California coast. Roads that added up to a thousand miles of cruelty, making the competitions set forth by Tony Tether and Sal Fish look laughable. At first, Thrun griped that it was impossible. If they'd just allow him one or two human interventions per route, it might be doable, he said. Page and Brin, echoing the DARPA director's thinking from 2003, refused. *It's only worth doing if you can take the human out of the picture altogether.* The team called this new challenge the Larry 1K. They had two years to reach the finish line.

On a Monday morning in early February of 2009, Chris Urmson found himself standing next to Dmitri Dolgov, each waiting in line to get his new employee badge. The men knew each other mostly by reputation. The Russian-born Dolgov had earned a PhD in computer science from the University of Michigan and, as a postdoc under Sebastian Thrun, had become a key member of Stanford's Urban Challenge team. Now the engineers were on the same side, along with a murderer's row of old rivals and new teammates. Thrun brought along his old Stanford allies and new contacts he'd made at Google; Urmson came in with his fellow CMU alum Nathaniel Fairfield. All together, these eleven men accounted for some of the best young engineering talent in the country. (Indeed, the world: The majority were foreign-born.) Thrun,

Urmson, Levandowski, Mike Montemerlo, and Dirk Haehnel—all of them DARPA Challenge veterans—were the official cofounders of the team, with particularly rich incentives. They believed that they had the skills to expand what they had done in the Urban Challenge to the real world. Very much a secret effort outside of Google (and even within it), they adopted a code name: Project Chauffeur.

Their effort to create what they termed the "self-driving car"—replacing the more common "autonomous vehicle" of the DARPA Challenge era—started by licensing the rights to the code Stanford's team had used in the Urban Challenge. It wouldn't be anywhere near good enough for the task at hand, but it would give the team what Urmson called a "jogging start." Instead of starting from scratch, they had something to improve. And the engineers tackling different aspects of the robot—perception, planning, and so on—could all dive in at once. It worked because they would follow the basic formula for a robot that the DARPA Challenges had put forward, with a drive-by-wire system implementing commands from a computer that saw the world with radar, cameras, and a roof-mounted, spinning Lidar system developed by Dave Hall.

The engineers set up in Building 1950, a curving glass structure on the eastern edge of Google's campus that also held Larry Page's office. Mapping whiz Montemerlo—who kept the roughly life-sized eagle trophy for Stanford's second-place finish in the Urban Challenge on his desk—took on the code that would teach the car to locate itself in the world. Dolgov developed the motion control system. Street View veteran Jiajun Zhu did perception; Fairfield worked on traffic light detection. Urmson ran the show from day to day, but as with many small teams tackling huge problems, Chauffeur had a tendency toward egalitarianism.

Life at Google had little in common with the research universities from which most of the Chauffeur team came. The money was better. For those coming from Pittsburgh, so was the weather. Lunch was free, as were breakfast, dinner, and snacks. Employees had access to gyms, workout classes, and massages. More importantly, Google offered the technological infrastructure to deal with huge amounts of data and code. But the thing that most struck Urmson was how empowered his team was. He'd spent most of his professional life in universities and was used to research grants with persnickety rules governing details like which pots of money could go to which equipment. Here, if he needed a new GPU, Lidar scanner, or car, he just ordered one. Meanwhile, Urmson settled into life in the Bay Area, where palm trees jockeyed for space with redwoods. He made the move a few months before his wife and two young sons, and in the interim accepted a generous offer from one of his new teammates: Anthony Levandowski knew Urmson casually, and offered him a spare room in his Palo Alto house.

One of the few team members without a PhD, Levandowski worked on hardware, as he had on Google Street View. He went to a Toyota dealership in San Jose, bought half a dozen Priuses, and retrofitted them to accept a digital overlord. Unlike with the car that crossed the Bay Bridge for *Prototype This!*, he figured out how to properly tap into the brake electronically, along with the steering and accelerator. Then he attached the sensors that would be their guides. A radar behind the front bumper watched the road ahead. One camera faced forward, another scanned 360 degrees. And atop a roof rack sat one of Dave Hall's spinning Velodyne Lidars. The team gave the cars names drawn from pop culture like KITT and KARR from David Hasselhoff's *Knight Rider*, and Car 54 (a reference to a 1960s sitcom about New York City cops).

As the cars came together, Levandowski bounced between Google and 510 Systems. His own company—most everyone on the team knew he owned it, and that the higher-ups knew too—served as a contractor, doing some of the hardware work. Levandowski moved the cars between Mountain View and Berkeley, even storing various parts in his own garage. And as ever, he delivered at remarkable speed. In the DARPA Challenges, teams had typically taken months to get a vehicle up and running. Levandowski put half a dozen cars together in a couple of weeks. "If you need to blow up a dam in Germany, Anthony's your guy," said Isaac Taylor, Chauffeur's operations manager, referring to Barnes Wallis, the English engineer who invented the bouncing bomb RAF pilots used to destroy dams behind enemy lines. "The general who's going to win you World War II."

The differences between Levandowski and Urmson were clear from the beginning. The roommates got along well, but they came at the problem—at the world in general—from opposite mind-sets. Where Levandowski moved as a blur, Urmson was methodical. Near the start of the project, he told Thrun he needed to rewrite the car's core communication software. Thrun, who leaned toward the fast and scrappy side, thought the system they had was fine. Even if it crashed occasionally, stopping to rework it could slow progress. Urmson insisted, and took several weeks to do the work. The result was a stronger, more reliable system. "He was damn right," Thrun later admitted.

The team started off testing in the parking lot of the Shoreline Amphitheatre near Google HQ, where Stanford's Grand and Urban Challenge teams had worked, and at NASA's Moffett Field a few miles down the road (yet another old military base pressed into service). But conquering the Larry 1K meant taking the inherent risk of driving on public roads, those streets and highways filled with everyday humans, and without the benefit of the police escort that had made the *Prototype This!* run through San Francisco possible. After starting 2009 by testing in private—with a security guard posted at the entrance to the parking lot—establishing the car's ability to execute a computer's commands and understand Lidar data well enough to guess what was a car and what was a human, they took their first few cautious steps into the real world in the spring.

As a warmup to the nasty, hundred-mile routes that made up the Larry 1K, Chris Urmson started the car on Central Expressway. Running alongside train tracks, the Silicon Valley thoroughfare was four lanes wide, with a grassy median separating northbound from southbound traffic, a forty-five-mile-per-hour speed limit, scarce pedestrians and cyclists, and mile-long stretches between the traffic lights the car did not yet know how to interpret. It was a relatively simple learning environment.

Before the road could welcome the robo-cars, however, it had to be mapped. If the Urban Challenge had imparted two practical lessons, they were that Lidar was vital, and good maps made moving through a complex, dynamic world a whole lot easier. Even for the teammates who'd spent years on Google's mapping projects, this was cartography on a new scale. Driving their Lidar-equipped cars, they catalogued the precise location of every lane line and curb, the signs and the traffic lights, all of it accurate down to ten to fifteen centimeters. That precision offered two key benefits. First, it took some strain off the car's ability to perceive its surroundings in real time. With an a priori knowledge of where it would encounter stop signs, speed limits, and the like, it wouldn't steam through an intersection if the sensors missed the red octagon, or risk a ticket if they didn't spot the school zone speed limit. Second, it made it much easier to pinpoint the car's location. GPS was accurate to within a few yards at best, an unacceptable margin of error for driving in a populated world. If the Lidar said the car was ten feet from that speed limit sign, and twenty feet from that light pole, and the map told it exactly where those things were in space, it could do a quick spot of math and deduce its own location. This concept of mapping and localization is not unlike the way people move through their house in the dark. They reach for landmarks like the corner of the desk or the bedpost to check where they are, and arrive at the light switch without stubbing a toe.

Maps in hand, the Googlers sallied forth, and relearned another lesson from the DARPA Challenges: how much they could accomplish when they were allowed to focus on nothing but making a car drive itself. They had no priorities but the Larry 1K. The tiny team engendered no bureaucracy, no pointless meetings. They reported directly to Larry Page. "How beautiful it was not to do email every day," Thrun said. And as they went from the old Stanford code made for the controlled environment of the Urban Challenge to something that could tame the wilderness of real roads, they saw tangible progress from one day to the next. Chris Urmson and Dmitri Dolgov could take the car out in the morning, see it fail to make a given turn, make some code changes after lunch, and that evening watch the car nail that same move. It wouldn't do it perfectly. It wouldn't do it every time. But with each code change, it got better, more reliable, more human.

After those first runs on Central Expressway, the team decided it was time they started making real progress toward their goals. Their tech still wasn't good enough for the wicked roads of the Larry 1K, but it could take on the environment that would move them toward the other goal Thrun had negotiated for them, logging one hundred thousand autonomous miles. Driving on the highway required higher speeds, and a mistake could be deadly. But compared to surface streets, it was easy to navigate, with all the cars going in one direction and no intersections, cyclists, or pedestrians to worry about. They started with a sixty-five-mile loop of Bay Area highways that took about an hour.

Instead of putting engineers behind the wheel on a regular basis— their time was better spent in front of a computer—Chauffeur fleet operations manager Isaac Taylor hired third party contractors. For $25 an hour (plus overtime and expenses), they would work as "safety operators," a new term for the role of riding in the driver's seat while the car drove itself, ready to take control from the computer if a crash seemed possible. Without a model to emulate—no one had ever tested a fleet of robots this way—Taylor created a new kind of testing regimen. Operators would go two to a car. One sat ready to take control if necessary, his right hand always near the big red button that would sever the computer's control of the car. (Hitting either pedal or turning the wheel had the same effect.) The other rode shotgun with a laptop, looking at the graphical user interface Dolgov created to turn the car's software and sensor data into something a human could understand. The car automatically created its own record of what it was doing, which the human annotated, noting where it drifted from its lane, when it braked for no apparent reason, how it handled turns, and any time the driver disengaged the autonomous system.

New safety operators went through a weeks-long training that started with a week in the back seat of a car worked by two seasoned hands up front. They were to observe only. No talking, no risking distracting the operators from their jobs. The only thing they were allowed to say was "disengage"—if anyone said the word at any point, the driver took control. After training, the operator could graduate to the right front seat of one of the cars building the high-res maps that so helped the self-driving system. Their first rides in one of the autonomous cars would be with one of Taylor's most experienced drivers. Eventually, they'd be tested by Taylor, Urmson, or another high-ranking Googler, who'd watch them work the laptop and make sure they knew how to produce data the engineers could learn from.

Before getting behind the wheel, they'd go through a rigorous course that covered how the system worked and how to handle a vehicle that went out of control. On a regular basis, Taylor, a former amateur rally racing mechanic, would bring in pro driving instructors and run clinics for new drivers. Taking over the trusty Shoreline Amphitheatre parking lot, they'd go through the kind of training police get, learning to handle emergency lane changes, recover from skids, pull over safely with a blown tire, and so on.

Safety operators worked eight-hour shifts that typically involved six to seven hours of driving time. Before leaving base, they went through a carefully detailed checklist, making sure the sensors and software were running properly, the computers in the trunk were good, the taillights and headlights worked, the tags were accurate and up to date, the drivers were well rested, and more. They were encouraged to take breaks when they needed a moment's rest, and were free to expense any snacks they picked up along the way. By October of 2009, Taylor had built a team of twenty operators, along with a few folks doing basic maintenance on the cars.

As the fleet started logging serious miles, the engineers fell into their own rhythm. When the safety operators came back from a run and submitted their data log, a software engineer would pick out an issue that had come up multiple times—an awkward merge, say, or ping-ponging left and right within a lane. The engineer would go over the data, looking for what might explain the problem, then start working on the algorithm in question, writing new code when tweaking the existing stuff didn't do the trick. Once they thought they had a solution, the engineers would run it through a computer simulation, and then send the software to a car that ran on a closed course. Then it would go to a car plying public roads, with one of those most experienced safety operators. If the problem persisted, the engineers would try another approach. If it went away, they dropped their bucket back into the well of problems, pulled up the next issue, and started the process over again.

They broke up the coding work with regular games of foosball. Several times a day, the Googlers would assemble, two or three on each side of the table. They'd pull, push, and spin the handles, moving the plastic footballers through time and space. But even in this smallest, simplest facsimile of the real world, no one could make the ball do exactly he wanted—not every time, anyway.

As the team got deeper into the spring of 2009, Urmson decided it was about time they started going after their grand challenge, the Larry 1K. Of the ten one-hundred-mile routes, Urmson settled on starting with the one that went from Carmel-by-the-Sea on the Monterey Peninsula to the coastal town of Cambria, via California's famed Highway 1. The winding, narrow coastal road might spook a human driver, but as Stanley, Sandstorm, and H1ghlander had shown on the 2005 Grand Challenge's Beer Bottle Pass, robots don't mind such things. Rather, this route looked to be easiest of the ten, because it included just a handful of intersections and traffic lights. The bulk of the drive would be spent sticking to a lane and not veering into oncoming traffic.

Before that, though, the caution-minded Urmson set up another task: making the robot run a four-mile stretch of a Highway 101 front-age road, including a few traffic lights, ten times without the driver needing to take over. Once they'd done that, he was willing to send his engineers—no contracted drivers for the Larry 1K routes—down to Carmel to start trying the official route. After a mapping car had run the route and Dmitri Dolgov had driven it to preprogram the software for the best path to take, Urmson, Dolgov, and Levandowski climbed in the Prius. Dolgov had programmed the car to slow down very late when going into curves, a bit like a racecar driver. Looking at driving from the standpoint of physics, it made sense. But the engineers quickly realized it wasn't comfortable for them, especially on a winding coastal road. And if their car was going to someday carry passengers, it would have to drive more like an everyday human.

But they let it carry on, and the car stayed within its lane. All the while, Urmson's hand hovered over the red disengage button. It finally came down when they encountered construction that cut the road to a single lane. The car wasn't anywhere near capable of understanding the worker directing traffic, and they weren't going to chance anything. The trio went back to Carmel and gave it another shot, after Dolgov had tweaked how the car approached turns. By the time they reached the construction site, the road was back to two lanes, and there was nothing to stop them from reaching Cambria. They had covered 101 consecu-tive miles as passengers in a self-driving car. Exhausted, they celebrated with a beer and dinner, reflecting on what they had done, and what that might mean. "We're going to make billions," Levandowski said. After spending the night down south, they headed back to the office and placed an empty bottle of cheap Korbel champagne on a shelf. On the label, they had written the route and signed their names.

Next, they tackled a one-hundred-mile stretch of the El Camino Real, the road built by the Spanish to connect the missions they es-tablished along California's coast. Modernity had transformed the dirt footpath into a proper paved road, complete with busy intersections and, on the stretch that Page and Brin had selected, 237 traffic lights.

Those proved easier to handle than the bit of the route that sent them through downtown Palo Alto, where the car's struggles around pedestrians and diagonally parked cars forced Urmson to hit the disengage button again and again. But each failure gave them something specific to fix, turning a huge, abstract challenge of driving safely into a series of tangible problems to solve. Eventually, another empty bottle of champagne joined the first one on the shelf.

The team really hit its stride in 2010, knocking off Larry 1K routes that took them around Lake Tahoe, south to Los Angeles, up the I-5, and along Pebble Beach's famed 17-Mile Drive. Chauffeur's capabilities expanded to handle all sorts of driving scenarios, but still relied on a combination of persistence and luck to guide the car from the start to the end of any given route. The engineers would try over and over, usually making their attempts late at night when traffic was lighter, making endless adjustments to the software. "We attempted them many times until we got lucky," said Don Burnette, one of the engineers who joined the team as it slowly grew over that first year.

The system at this stage was capable, but its makers were specifically hand-tuning it for whatever route it was attempting at the time: how aggressive to be on a given merge, where exactly to be in the lane at every moment, what degree of color counted as red given the angle of a traffic light and the sun at a certain time of day. Burnette was in the car with Urmson, Dolgov, and Levandowski one night, late in the summer of 2010, when the resulting brittleness struck near the very end of a Larry 1K route. They were charged with traversing the five bridges that cross the San Francisco Bay, and had reached the final one—the Golden Gate—when they realized the toll booth for the lane they had selected was closed. The car's programming didn't offer the flexibility to jog to the left or right. Down went the red disengage button.

The next day, they tried again, and this time hit trouble in Tiburon, just before reaching the Golden Gate. They were on a narrow road, with cars parked on both sides, when another driver approached from the opposite direction, then pulled over to let them pass. The robot, though, was programmed to require more clearance than it had. Levandowski got out of the car and asked the driver to move farther over, without being able to say it was a robot car and that if they could just get past, they'd be closer to making Larry Page a happy man. The man was confused and annoyed, and Urmson wanted to defuse a potential problem. Down went the red disengage button. The trio returned to Mountain View, refueled the Prius, and tried yet again—this time making it all the way through. Returning to the office in the early hours of the morning, they put another bottle of Korbel on the shelf and celebrated with a game of foosball.

In the spring of 2010, the team had suffered its first serious crash. Two safety operators were doing a run on surface streets, testing out a new traffic light detection system. On Central Expressway, when a light turned yellow, the car hit the brakes, and was promptly rear-ended by a pickup truck. The robot, a maroon Prius named KITT, was totaled. The two safety operators—and their friend who was definitely not supposed to be sitting in the back seat—were fine, as was the truck's driver. Nobody called the cops, and the press didn't get wind that Google was testing self-driving cars on public roads, and crashing.

That crash came a few months after an incident that stemmed from Levandowski's endless desire to move ahead. As fleet operations manager, Isaac Taylor determined how the cars tested on the highway. While he was on paternity leave, Levandowski—the hardware lead—decided to expand the test envelope to new situations, including driving consistently in the right lane of the highway. A team member alerted Taylor, who returned to the office furious. After arguing, he and Levandowski took a Prius and a nearby truck onto the I-280 to see if Taylor was too cautious or Levandowski too aggressive. Perhaps because the robot's map didn't include subsequent on-ramps (part of the reason it was meant to stay to the left), the Prius didn't make room for a Camry merging aggressively onto the freeway. Levandowski moved to avoid a collision, and the Camry spun out. This event, too, went unreported (until 2018), although Urmson included it in a 2011 presentation to the Institute of Electrical and Electronics Engineers—minus the argument—saying the Camry's driver ended up with "a little bit of excitement in his day."

On September 27, Urmson, Dolgov, and their fellow engineer Andrew Chatham got into a Prius to try the tenth and final leg of the Larry 1K. This was a particularly nasty and diverse one, taking them through Palo Alto, up Skyline Boulevard into the Santa Cruz Mountains to the west of Silicon Valley, then down to the Pacific Coast on Highway 1. From there, the route went north into San Francisco, then east through the city. In the Russian Hill neighborhood, it went north and then south four times within a mere 450 feet. Page and Brin had taken great pleasure in picking the hardest roads in California, and they weren't going to miss San Francisco's famously crooked Lombard Street. Once again, the car showed that humans and robots have very different definitions of difficulty. A windy street that could stress a human didn't perturb the Googlers' software. The true test came when the Prius turned onto Market Street. This major thoroughfare cuts diagonally through the street grid, making for tricky intersections. It carries not just cars and trucks and taxis, but bikes, buses, and a streetcar line. Its wide sidewalks teem with pedestrians. But that Monday, the car handled it all. With the rest of the Chauffeur team tracking the Prius's progress on monitors in the office, Urmson, Dolgov, and Chatham let the car take them into the hills at San Francisco's center. Then, high above the city they had just navigated, they reached the end of Market Street, and the end of the Larry 1K.

The moment called for more than a single bottle of champagne. The team, safety operators included, gathered for a grand party at Sebastian Thrun's house in the hills of Palo Alto. They threw one another into the pool and played "human foosball" on an inflatable field. The celebration was well deserved. This team of all-star engineers had taken on a challenge that made Tony Tether's DARPA races look simplistic, and they had conquered it in well under the allotted two years. Almost impossibly quickly.

By another measure, they had finished just in time. A few weeks before they completed the final run, Thrun had learned their cover was blown. John Markoff, the *New York Times* tech reporter who had been in Stanley when the Grand Challenge champion crashed into a bush, had gotten a tip from the friend of a Google safety driver that the company was testing robot cars on public roads. Knowing a story was coming out no matter what, Thrun tried to ensure it made Google seem innovative, not reckless. He invited Markoff to climb into one of Chauffeur's Priuses for an autonomous ride down the freeway and through Mountain View, with Urmson behind the wheel. While the team made clear that it had years of work to do before it could commercialize the tech, and still struggled with things like traffic cops' hand gestures, its system impressed the reporter. "The Google research program using artificial intelligence to revolutionize the automobile is proof that the company's ambitions reach beyond the search engine business," he wrote in the October 9, 2010, article that introduced the world to Google's self-driving car.

— 8 —

The Un-Car People

IN 2011, GOOGLE WORKERS LOOKING TO GET ACROSS THE COM-
pany's sprawling Mountain View campus had a few options. They
could drive, if they had a car and happened to be moving in the few
precious hours between the end of the morning rush hour and the
start of the evening crush home. They could walk. Or they could hop
on one of the colorful bikes Google stocked around the campus for
people to use as they liked. But soon after the *New York Times* exposed
the self-driving car project, the Chauffeur team started testing a new
way. It attached its sensors to a small fleet of electric golf carts and sent
them onto the sidewalks that crisscrossed the campus, puttering along
and yielding to pedestrians. This project, dubbed "Caddy," was meant
to see what a low-cost version of their tech could do. The answer was
not much, but the posing of the question represented a new stage for
the Chauffeur project.

Mastering Market Street had brought up the new challenge of mastering the market. By completing the Larry 1K, the team, still just a
dozen or so strong, had proven that its car could handle an array of difficult routes at least one time, usually after a series of failed attempts.
The system was far from what Red Whittaker would call "rock solid."
True success required making a car that could handle everything, every
time. It required generalizing a technology that worked in very specific

conditions. And it raised questions that hadn't mattered in the race to complete those thousand miles.

Questions like: How much more time and money would it take to finish their work? What did "finished" even look like? How good was good enough, and how would they know that their cars were safe? What would regulators think of this whole idea? How about insurance companies? How about the public? What kind of product were they making, anyway? Should Google start manufacturing cars, partner with existing automakers, or produce an aftermarket solution that individual drivers could have installed on their vehicles?

The limitations of the technology hemmed in the paths to commercialization. The cars needed incredibly detailed maps, created by Lidar laser scanning, that described their environment down to a few centimeters. Even Google couldn't do that for the whole country, let alone the world, not with high-definition maps that had to be updated constantly, accounting for things like shifted lane lines and construction zones. Moreover, driving habits, rules, and conventions changed from one place to another. Humans handle this without too much trouble. Minnesota drivers tap into their aggressive side to get along in Boston; a tourist clueless enough to rent a car in New York learns from an aggrieved pedestrian that the city forbids right turns on red. But as with so many things humans do naturally, it wasn't clear that a robot designed to drive in one region could easily adapt to another. At this stage, Chris Urmson and his teammates were still teaching the cars how to handle routes one at a time.

Put together, these factors made clear to the Googlers that if training a self-driving car to handle all roads—a prerequisite for any vehicle looking to replace the traditional, human-driven type—wasn't impossible, it could take decades to accomplish. The trick, then, was to limit the scope of the problem. This is the concept known as an operational design domain, which limits what a technology has to do by limiting the challenges it will face. To create a winnable race, Tony Tether had excluded dynamic obstacles from the first two Grand Challenges.

He kept red lights and pedestrians out of the Urban Challenge. The Chauffeur team wouldn't have that kind of control over their environment, but they could be picky in selecting their missions. One of the first promising possibilities was to restrict their technology to one kind of road.

The United States contained more than 164,000 miles of highway. A lot, sure, but not so many Google couldn't map them. And compared to nosing through San Francisco or even suburban Mountain View, highway driving was easy. The vehicles moved in the same direction and didn't encounter traffic lights, pedestrians, cyclists, or four-way stops. Cruising around Bay Area interstates, Chauffeur's cars had zoomed past the one-hundred-thousand-mile marker long before completing the Larry 1K. Anthony Levandowski was commuting from Berkeley in a Chauffeur car. Sebastian Thrun thought it was working great. He pushed the team to package a version of its software and hardware, which automakers could build into their standard cars and sell as a luxury add-on. On surface streets, the human owner would drive, just as she normally does. But on the highway, she'd have the option to let the car do the work. It sounded reasonable, provided they could rein in the costs and make the sensors reliable. As their technology continued to mature, they could offer increasingly capable systems, and sometime in the future get to surface streets.

Completing the Larry 1K had made another thing clear: These engineers had created something amazing, a technology that could change the world. But they didn't own it. Google did. Nice bonuses and good salaries were measly compared to what their employer stood to make from a viable self-driving business, whatever that would look like. In 2011, led by Levandowski, some of the engineers indicated to their bosses that they were interested in leaving and launching their own effort, where they would dictate the terms and reap the profits. To keep them in-house, the Google higher-ups created a compensation plan that worked much like that of a startup. Along with their salaries, the engineers would receive bonuses calculated as percentages of the Chauffeur project's value. Those payouts would come after four, eight, and twelve years, so they had to stick around at least until late 2015 to see serious money. But if they succeeded, it could be serious indeed.

With dollar signs in their eyes, the Chauffeur team charged ahead. In mid-2011, they added a new weapon to the arsenal, Dave Ferguson, the New Zealand native who had worked on Carnegie Mellon's Urban Challenge effort, teaching Boss how to get from one waypoint to another. Once settled in Silicon Valley, he led some of the program's earliest work in machine learning, the technique that gave a computer lessons instead of laws. It was the same kind of approach that CMU's Dean Pomerleau and Todd Jochem had used to make their No Hands Across America trip in 1995, teaching their car to pick out lane lines by showing it lots of images of lane lines.

With the cars still spending lots of time on the highway, Ferguson and his teammates started developing a way to detect when other cars were going to change lanes. But like a good human driver, their cars couldn't rely on turn signals—too many people neglected to use them, or didn't turn them off after moving. To use machine learning to make a computer recognize a cat, one doesn't tell it to look for paws, a tail, and so on. One shows it photos of cats until it catches on to the pattern. Ferguson and his team did the same thing, feeding their software tons of examples of cars changing lanes, and tons more of cars not changing lanes. Contained in those examples were all the nuances an experienced driver knows to look for: for example, the way someone moves within his lane that makes you pay extra attention. Google's cars didn't ascribe logic to their predictions the way humans did—*that guy in the BMW is going to cut me off because he's trying to get out from behind that truck*—but they learned to look for the same kinds of clues. The team would go on to apply this way of predicting behavior to all sorts of situations and actors, including cyclists and pedestrians. They used similar techniques to teach the car to spot construction zones, and places where reality didn't match what was in their preloaded maps. Machine learning made recognizing traffic signs far easier, saving programmers months of work spelling out what a stop sign looked like from every angle in every type of light.

Up until this point, machine learning had not been an especially popular tool in computer science. That changed in 2012, when University of Toronto researchers created a newly powerful kind of neural network, the programming that underpinned the learning process. Benefiting from ever faster computers, this system was a milestone that woke many up to the potential of a new type of machine learning, called deep learning. Over the next few years, computer scientists used deep learning to defeat human champions at the ancient and complex game of Go, train voice recognition software, help Facebook put names to the faces in photos uploaded by its users, and much more. After their breakthrough, the Toronto researchers moved to Mountain View to work on a new artificial intelligence effort called Google Brain. For one project, they teamed up with Ferguson's crew to improve the way the cars spotted people on foot. In a few months, they made the software a hundred times less likely to miss a pedestrian. For the Chauffeur team, access to such help validated the choice to stick with Google.

In early 2011, soon after the completion of the Larry 1K, Anthony Levandowski set his sights on a different problem. Always eager to accelerate, he wanted to see his robots on the road, and not just as prototypes. Chauffeur's tech wasn't anywhere near ready for commercial application, but when it was, California didn't look like the best place for it. Testing the tech was legal by virtue of not being explicitly illegal, but it was only a matter of time before the Golden State brought its traditionally heavy regulatory hand to bear, especially now that the self-driving program was public knowledge. When it did, who knew what rules might box in—or even forbid—what Google was spending so much time and effort to build?

Levandowski found a way to get ahead of the problem at the Consumer Electronics Show in Las Vegas. He was working Google's booth, showing off Chauffeur's car, when he met David Goldwater, a former Nevada state legislator turned lobbyist. Goldwater suggested Google start the legal push by focusing on a smaller, more permissive state— his. Working out of Goldwater's office, the two drew up a bill that would make it explicitly legal to operate autonomous vehicles on Nevada roads and direct the state's Department of Motor Vehicles to create appropriate rules. They had to move quickly: After the Nevada legislature ended its session in June, it wouldn't meet again until early 2013.

In early April, Goldwater, representing Google, went before the State Assembly Committee on Transportation, bill in hand. "This is a great opportunity for economic development," he said. "Getting the law ahead of technology allows us to potentially attract manufacturing, engineering, and development aspects of this kind of technology." It was only after the bill was introduced that Google's public affairs team found out about it, and called Sebastian Thrun. Thrun scolded Levandowski and told him to withdraw the rogue bill. Levandowski ignored him. In June, the bill sailed through both houses of the legislature. The chance to lure in this kind of technology had obvious upsides, especially since Nevada's real estate– and tourism-focused economy was still suffering the effects of the 2008 financial crisis.

Levandowski's initiative paid off. After the bill became law, the Chauffeur team and Google lawyers worked with the Nevada DMV to craft rules that wouldn't hamper the effort or let anyone's robo-cars rampage about. DMV chief Bruce Breslow didn't want to burden any company with many regulations, but he did want the cars to indicate whether they were in autonomous or human-driven mode, the way a taxi light makes clear whether the driver is on or off duty. If one crashed, he reasoned, the police should know if it was empty at the time, or if there might be an injured driver staggering through a field nearby. But Google didn't like the idea. Lawyer David Estrada argued that people had a habit of driving dangerously around vehicles they knew were autonomous, almost bullying the overly cautious robots. And Google had the trump card: It didn't have to test in Nevada. "If we did something as a state, one little extra thing, and ten other states didn't do that, they wouldn't come to our state," Breslow said. The indicator light, which his colleagues dubbed the "Bresbulb," died a quick death.

Just eight months after the bill's passage, the DMV published its rules. They required Google (or any other self-driving developer) to equip the car with a device that would work like the black box on a plane, logging the sensor data from any crash. The company would have to keep two drivers in the vehicle, get approval for how it trained them, and put up a hefty surety bond. It would also face a driver's exam. The car would have to drive competently on a few surface streets, merge onto and exit the freeway, then enter a school zone and adjust its speed accordingly. For a final flourish, the DMV testers would send the car down the busiest road in the state.

On May 1, 2012, a Chauffeur Prius rolled into Las Vegas, with Chris Urmson in the driver's seat and Anthony Levandowski riding shotgun. In the culmination of a fourteen-mile test drive, it passed Caesars Palace and the Flamingo Hotel, smoothly navigating through intersections managed by arrays of traffic lights, all of it previously mapped in careful detail. The car paid attention to the throngs of pedestrians but ignored the city's hokey glitz, noting the truck ahead of it but not the ad on its back for "HOT BABES." Breslow sat in the back seat, impressed but unaware that his test was finally delivering on Tony Tether's initial vision for the first Grand Challenge before practical considerations forced DARPA to settle for a race through the desert: seeing an autonomous vehicle drive itself down the Las Vegas Strip.

After getting the official thumbs-up, Chris Urmson smiled as he attached a new license plate to the robot's bumper. Bright red, it carried a figure eight–shaped infinity symbol and read "AU 001." Google had history's first legally sanctioned self-driving car.

That car and its brethren, though, never did much testing in the Silver State. The real upshot of the Nevada legislation was giving Google's home state a kick in the pants. In California, a state senator named

Alex Padilla used Nevada's legislative scoop to get his own state to act. The Golden State was the land of cars and of technology, he argued; it had to be a leader here. And it couldn't risk seeing major employers like Google moving jobs across state lines. California settled on stricter rules than those in Nevada, requiring that companies testing autonomous tech publicly report any crashes, and file annual reports logging how many miles their cars drove and how many times their safety drivers took control back from the robots, a metric called "disengagements." But it moved quickly: In September 2012, governor Jerry Brown arrived at Google's headquarters in what was still called the driver's seat of an autonomous Prius, with Sergey Brin beside him, and signed his bill into law. Other states followed suit. Later that year, Florida and Washington, DC, created similar legislative frameworks for regulating driverless cars. Lawmakers in Arizona, Hawaii, Oklahoma, and New Jersey all considered similar rules in 2012, but failed to pass the bills before their legislative sessions adjourned. By the end of the decade, more than half the states would legalize self-driving testing, either through legislative acts or executive orders. Congress even passed a bill that would set national rules, but it died in the Senate, held up by a few legislators nervous about the implications of loosing robots on the public.

Levandowski's unauthorized push for rule-making may have kicked off a national frenzy for driverless cars, but he was having trouble on the home front. In early 2011, the relationship between Google and his company, 510 Systems, had risked becoming problematic. Sebastian Thrun knew that Levandowski had authorized Chauffeur to buy hardware from his own company. But 510 and its sister company, Anthony's Robots, were developing their own self-driving software and a proprietary Lidar scanner. Cock your head just right, and Levandowski looked like a potential competitor to the Google effort he had helped start. And Topcon, 510's biggest customer, was hoping to acquire the company, along with all of its intellectual property.

By February 2011, the Google chieftains had decided to make sure no other company got their hands on Levandowski's work. They moved to buy 510 Systems and Anthony's Robots, but Levandowski's questionable setup raised the eyebrows of David Lawee, the exec who led Google's acquisitions team. "I don't understand this guy at all," Lawee told Thrun in an email. "As Google, I suppose I'm prepared to take the risk with Anthony, but I can say definitively that if I was choosing a business partner to start a company with, there is no way in hell that I would proceed."

Thrun was having his own doubts about Levandowski. He was starting to move away from his leadership of the self-driving team, and had considered putting the young engineer in charge. But he soon dropped the idea, replying to Lawee later that day: "I now question whether Anthony is fit to lead Chauffeur. I hear similar concerns from the engineering team. Several people there have contacted me that they have concerns about Anthony's commitment and integrity." Some team members disliked the fact that Levandowski owned a company from which they were buying hardware—it just seemed shady. He didn't help his reputation by wearing a custom T-shirt reading "I Drink Your Milkshake." It was a reference to the 2007 film *There Will Be Blood*, about a conniving oilman who bilks various people out of their fortunes, murdering a couple along the way. Others at Google didn't care, figuring that if Levandowski was getting away with it, good for him. "He plays above board," said one teammate. "He just changes the rules as he goes."

Levandowski, though, wanted the top spot, and Thrun wasn't prepared to do that. "If he is the single leader, a good number of team members will leave," he told Larry Page in an email. But Google went ahead with the 510 deal, eager to keep Levandowski's dynamism, as well as his intellectual property, in house.

In late April, the fifty or so employees of 510 Systems got an email with the subject line, "Invitation: Mandatory All Hands Meeting—BE ON TIME," meaning 10 a.m. the next day. When the group gathered, Levandowski announced that Google had offered to acquire the company and make them part of its self-driving car effort. Some liked the idea.

Others thought they were selling themselves short, that they had created a valuable tool for mapping and should pursue that as a business. Memories differ on whether the group voted on the decision and whether those ballots, if they existed, influenced Levandowski's decision. One way or another, the deal was done, and Levandowski benefited more than anyone. Of the roughly $20 million Google paid for 510 Systems, he took about half. His cofounders Pierre-Yves Droz and Andrew Schultz split the remainder, and some money went to employees as bonuses. Workers who had joined 510 in its first few years and expected a big payout—standard fare for an acquired startup—were disappointed. Levandowski had sold 510 for just enough so that the real money stayed with the founders— anything above the $20 million mark would have gone to the employees.

Soon after that meeting, the 510 crew boarded a bus that took them to Mountain View, where they joined the Chauffeur engineers—their new teammates—for a rather awkward picnic. About half the 510ers were offered jobs at Google, including the six-person Lidar team. They would get shares of the bonus setup the Chauffeur team had accepted in exchange for staying put at Google. The other half were expected to wrap up 510's existing business with companies like TopCon. At his exit interview, Kevin Rauwolf, one of the engineers not going to Google, vented over being forced to compete with colleagues for a job. He told Levandowski he hoped he'd never get another chance to treat people the way he'd treated them at 510. Levandowski, though, seemed not to register the emotion. "Hope I get to work with you again," he said cheerily, as Rauwolf walked out.

The millions that Levandowski reaped from the acquisition were just the amuse-bouche. When it came to portioning out shares of Chauffeur's equity in the new bonus structure, most of the team's key members would get a .5 percent share. But Levandowski, partly to compensate him for selling his companies for a relatively low price, would get a tenth of the entire pot, twenty times more than his colleagues. The 10 percent, according to Thrun, was based on a directive from Page, to "make Anthony rich if Chauffeur succeeds."

———

Success depended on Chauffeur's ability to put a product into the world, to put people in self-driving cars. The question of how to do that, exactly, triggered a lengthy process that would ultimately erode the bonds that kept the team together.

In the spring and summer of 2011, Chauffeur hired its first wave of non-engineers, charged with researching the best way to commercialize self-driving technology. From the start, the team was focused on a system that would work on the highway. Someday, their tech might be good enough to let nondrivers—the blind, the elderly, the young—use a car on their own, and to make a truly world-changing product. But for now, this limited application made sense. The cars worked well in the simple, one-way environment, and considering how many drivers shelled out extra cash for rudimentary cruise control, it could be a valuable proposition.

The product team of just a few people started by interviewing Google employees who owned Tesla Model S sedans and Roadster sports cars, figuring there'd be overlap between fans of the high-tech electric car maker and fans of what they were designing. They hacked together a simulator, using projectors to make the windshield of a Lexus SUV display the view of the drive from Mountain View to San Francisco. They put Google employees inside, to see what they'd do inside a car they didn't have to drive. And from surveys, they realized that many of their well-paid colleagues drove cheap cars—seemingly unconvinced by the lifestyle that luxury brands like Cadillac, BMW, Audi, and Mercedes-Benz were selling. These "un-car people" didn't care about the vehicle itself. But they might well shell out extra cash for a system that let them spend more time on the road relaxing or talking to their kids, even if it was just on the highway. The average American worker spent nearly an hour commuting by car every day. In the traffic-choked Bay Area, it could be much longer.

With this sort of idea in mind, Chris Urmson and Anthony Levandowski had started making trips to Detroit and Germany in late 2010 and early 2011, to explore partnerships with established automakers. Far from intrigued, the executives they met with were either dismissive or uninterested. They considered testing on public roads reckless, thought the roof-mounted Lidar looked impossibly stupid, and said no, they were not interested in working together.

As this research moved along, the engineers designing the underlying technology staked out their own positions on how Chauffeur should move ahead. Levandowski, always looking to charge ahead, advocated for selling an aftermarket kit that could make any car drive itself on the highway. Urmson wanted something more carefully integrated into a car, believing it would be more reliable, even if it took longer to produce. The two men were no longer roommates; Urmson had moved out when his wife and sons arrived from Pittsburgh, and had put a down payment on a house with his bonus from completing the Larry 1K. But they were finding it hard even to be colleagues.

The pursuit of the Larry 1K had preempted potential tensions by motivating everyone toward a shared, well-defined goal. Now the differences in how they viewed Chauffeur's future, and the path to get there, became a source of conflict, and part of a bigger power struggle: Both Urmson and Levandowski made it clear that they wanted to lead the project, and that if the other had too much power, the program could be ruined. Thrun thought each offered a valuable skill set, and that differences in perspective were healthy. But this was not Abraham Lincoln's "team of rivals," overcoming their differences to help preserve the Union and win the Civil War. It was more like that war itself, where every topic of discussion could be turned into a debate, an argument, a swear-laden shouting match.

The team managed to find casus belli in the least important places, like what sorts of buttons the car should have. Some wanted two buttons: green to engage the system, red to disengage. Others thought it would be simpler to have just one for both on and off. They argued over whether they should add new buttons, or repurpose those already in the vehicle. They argued over how to display the speed the driver set for the vehicle, whether as the absolute number (say 70 mph) or as an offset to the speed limit (65 + 5). These would have been valid debates for a team that was on the verge of putting a product in the market. But Chauffeur wasn't anywhere near that point. The engineers—none of whom had relevant product, design, or business experience—were fighting over the design for their new fleet of Lexus SUVs (a more comfortable, capable option than the original Priuses). A design that would be seen not by customers, but by their own safety operators.

"We argued about the dumbest things and wasted so much time," said engineer Don Burnette. "How impactful or meaningful was that button discussion in retrospect? It was a complete waste of time."

A perhaps more important consideration was how people would feel about the very concept of a self-driving car. In 2011, the Chauffeur team tried putting people new to the technology in the driver's seat, attaching biometric sensors to their chest, finger, and wrist. One new employee who tried it out was directed to turn on the autonomous system while speeding down the 101 freeway. His stress levels, he was told, matched those of a person waiting to be executed.

Tensions between Urmson and Levandowski didn't affect the team on a daily basis, but the two occasionally erupted at each other, forcing Thrun to intervene. In an email, he admonished the team for in-fighting. They had climbed one mountain, he said, but were standing at the base of a much taller one: "We have not yet saved a single life. We have not yet enabled a single blind or disabled person to operate a car."

Thrun didn't want to fire either of the engineers, and neither wanted to leave: The Chauffeur equity plan they agreed to in 2011 stipulated they would be paid their bonuses—which promised to be in the seven-figure range—only if they stuck around until 2015.

While Chauffeur's engineers bickered over how to bring their tech to the highway, a new movement was changing how people moved around cities, one that might offer a different path for self-driving cars to enter the world.

One herald of this change was Larry Burns, the General Motors R&D chief who had greenlit the automaker's sponsorship of Carnegie Mellon's Urban Challenge bid. Burns had spent the first decade of the millennium pushing new ways to think about the auto business, including powering cars with batteries and hydrogen. While GM and its competitors fought to stick two SUVs in every American driveway, he saw the personal car as a waste of space and power, especially in increasingly congested cities. In 2008, Burns led the development of Project PUMA, for Personal Urban Mobility and Accessibility. The two-wheeled electric concept looked like the back half of a rickshaw. Made in collaboration with Segway, the PUMA had room for two people and could go thirty-five miles between charging breaks. Burns called it a "mobility pod," designed for moving around cities in a space- and energy-efficient manner. He hoped it would prove that GM was open to a changing future, but by the time he showed it to the public in 2009, the company had bigger issues to worry about: It was headed into bankruptcy.

Burns retired from GM soon after, but kept with the idea. In 2010, he published *Reinventing the Automobile: Personal Urban Mobility for the 21st Century*. In the book, he and his coauthors pitched a vision of a world in which emerging technological trends allowed for cars that didn't take up unnecessary real estate, didn't belch greenhouse gases, and didn't sit idle nearly all day. They imagined vehicles that ran on clean electricity, communicated with one another, were small yet safe and comfortable, and were shared instead of privately owned. Maybe, someday, they'd even drive themselves, making them available to anyone with or without a driver's license. "When effectively combined," they wrote, "the ideas behind this reinvention promise to enhance our freedoms and stimulate economic growth and prosperity while eliminating many, if not all, of the negative side effects of today's automobile transportation system."

General Motors didn't exactly buy into this idea, but in January

2011, Burns took a job with a company that might. As Google's Chauffeur team inched toward commercializing its tech, Chris Urmson asked Burns, whom he knew from the Urban Challenge, to help them navigate the foreign world of the auto industry. Burns gladly signed on as a consultant, and took the opportunity to give the team a presentation on the ideas he'd laid out in *Reinventing the Automobile*. But like any revolution, this one wouldn't follow the enlightened writers' playbook. And the vanguard was already in the streets.

In 2009, San Franciscans trying to flag down one of the taxis plying the streets of the City by the Bay faced a challenge. San Francisco allowed a measly fifteen hundred cabs on its streets, and drivers who had spent their savings on a medallion mostly earned the money back by sticking to the dense downtown and most profitable routes, shuttling business travelers between the big hotels and the airport fifteen miles to the south.

It was a point of frustration for Garrett Camp, a Calgary native who in 2007 sold his website, StumbleUpon, to eBay for $75 million. He wanted to enjoy his money in San Francisco, but often found himself unable to find a cab when he wanted to go out or go home. He ended up relying on the black cars roaming the city, and that's where he got his next idea: It would be great, Camp thought, if you could use your iPhone to have a car come to you, whenever and wherever you wanted one, without having to ring up drivers until you found one who was free. It would be awesome, or to use the German term that came to his mind, *über*.

UberCab launched in San Francisco in May 2010, just as the Chauffeur team was moving into the final stages of conquering the Larry 1K. Later that year, Camp's friend and Uber investor Travis Kalanick took over as CEO. In the brash, aggressive style for which he would soon be both revered and reviled, he started bringing the service to cities across the country and around the world. In these early years, Uber was a relatively small business, used by people looking for luxury rides at discount prices. That changed in January 2013, when Kalanick followed the lead of a new rival called Lyft and expanded beyond the lightly regulated black car market with a new service. Instead of connecting users to drivers with commercial licenses, "UberX" would send their ride request to virtually anyone with a car, for a reduced fare.

This idea, known as ridesharing or ridehailing, is what made Uber and Lyft significant. Companies before them had used phone apps to connect riders with taxis. As the ridehailing companies spread to one city after another, regulations decreeing how many people could drive taxis and how much they could charge were steadily rendered meaningless. Within a few years, some forty-five thousand Uber and Lyft drivers were operating in San Francisco, a thirtyfold increase over the number of taxis (though many drivers worked part-time). While Lyft stayed domestic, Uber went international, eventually launching its service in hundreds of cities on six continents.

Debates raged over whether the services delivered on their lofty promises to limit traffic or actually made it worse; if "gig economy" drivers should be treated as employees with worker protections and benefits; if a background check was enough to keep passengers safe. In Paris, taxi drivers went on strike, blocking streets and lighting tires on fire in protest. Uber continued operating in the city. When Austin passed a law demanding drivers be fingerprinted, Uber and Lyft left in protest. They returned a year later, after successfully pushing for a laxer state bill that superseded the city's regulation. When New York City mayor Bill de Blasio proposed capping the growth of ridehailing fleets, Uber showed users a tongue-in-cheek "de Blasio" mode, displaying twenty-five-minute wait times for cars. The mayor backed down. (In 2018, the city did enact a limit on ridehail vehicles.) Soon, Uber was providing millions of trips a day. In 2013, a New York City taxi medallion was valued at more than $1 million. Six years later, one could be had for $138,000. Why pay for the right to drive a taxi when one could use one's own car and drive for Uber or Lyft? Or for Via, Gett, DiDi, Ola, Careem, or any of the other ridehailing companies that had popped up around the world?

From one perspective, ridehailing fit perfectly into Larry Burns's vision of a future with fewer, shared cars. Lyft cofounder John Zimmer hit the utopian note especially hard, presenting his company as a way to end personal car ownership. "Cities of the future must be built around people, not vehicles," he wrote in 2016. "By 2025, owning a car will go the way of the DVD." From another perspective, ridehailing debased the rule of law, increased traffic, stole fares from public transit, punished taxi drivers who had invested in commercial licenses and medallions, and locked drivers into a system that paid little and provided no benefits or safety net.

Either way, Uber's valuation skyrocketed from $3.5 billion in early 2013 to $82.4 billion when it went public in 2019. But valuations didn't mean profits. As they raced to expand and undercut each other in search of market share, Uber and Lyft racked up massive losses, sometimes more than $1 billion per quarter. They constantly experimented with pricing schemes to balance supply and demand. But for all their tweaks, they were stuck facing a fundamental conflict: Higher fares drew in drivers; lower fares attracted passengers. They needed a lot of each to succeed.

Travis Kalanick got his first taste of the solution to that problem one morning in August of 2013, when he stepped out of the Four Seasons Hotel in Palo Alto and saw a Google driverless car waiting for him, the Lidar spinning away on its roof. With its ridehailing business taking off, Uber was a hot property. Larry Page was interested in investing some of Google's cash reserves in the startup, and he had invited Kalanick over for a meeting. Sending Chauffeur's chariot was part of the VIP treatment. Kalanick wasn't too impressed with the robot's driving. "The car barely worked," he said. "At times it would do some things it was supposed to. And other times it wouldn't. It was just not there yet." But it didn't take him long to imagine what a working self-driving car could do for his business. "The minute your car becomes real, I can take the dude out of the front seat," he told Google Ventures partner David Krane. The dude who pocketed about three-quarters of each Uber fare.

With Google's self-driving cars, Kalanick could eliminate one of his company's great roadblocks to profitability.

And with a ridehail system like Uber's, Google could finally have a clear, ready-made path to commercializing its tech. Forget trying to produce some limited system for operating on the highway and linking up with shortsighted automakers. The Chauffeur team could choose a city, or part of one, to map in high-resolution and deploy its robots to scoop up Uber users. If a rider wanted to go somewhere Google hadn't mapped or that required complex navigation, the dispatcher would send a human driver. As the cars got better and Chauffeur figured out the business side of the scheme, they'd take on more routes, then more neighborhoods, then more cities, gradually making the humans irrelevant. That idea of a mixed fleet would allow for the tech to emerge gradually, and to start making money before it was perfect.

In that meeting, though, the Google contingent didn't make any promises on the self-driving front. It talked up how the tech giant's resources could help Uber grow—and produce valuable returns on a Google investment. Page put $258 million into Uber and sat Google exec David Drummond on the startup's board of directors. Kalanick cashed the check and figured Google would come back to discuss an autonomy collaboration. He didn't know that in a garage on the outskirts of the Google campus, the Chauffeur team was already developing a product not to run on the Uber network, but to challenge it head-on.

Through 2012, the Chauffeur team had been focused on how to produce a self-driving system for the highway, something built into a car a human would drive most of the time. The idea seemed sensible. A taxi-type service was intriguing, but driving on surface streets was extremely tricky, and the legalities of totally driverless cars were hazy. "It looked like step number two, because it was so much harder," Thrun said. If the big automakers weren't interested in collaborating, Google could find a willing partner in China. Or maybe in California: Google was considering buying Tesla at the time. Elon Musk's electric automaker was on the verge of bankruptcy, and Musk and Page had discussed a deal. (That ended up falling through, and Tesla managed to stay in the black.)

So in early 2013, the team brought in their first guinea pigs, giving one hundred Google employees self-driving prototypes to use in their daily lives. These beta testers got two hours of training. They were told to use the self-driving function on the highway only, and to keep their eyes on the road at all times, staying ready to take control back from the computer. The cars were good, but far from perfect, and nowhere near ready to be left on their own. The team rigged up cameras inside the cars and waited for the results to come in.

Most of the drivers went through the same process: They'd start off nervous, keeping their hands on the wheel and right foot just over the brake pedal. Once the car started driving well, they'd be amazed, and slowly release their grip on the wheel. Another few minutes, and they'd come to trust the computer. Soon, they would stop paying attention to the road. Chris Urmson was horrified to see them playing with their phones and applying makeup while going 65 mph down the freeway in prototype robots. One guy reached into the back seat, rooted through his backpack, pulled out his laptop, and put it on the passenger seat. Then he turned around again, dug around for a charging cord, and connected his phone to the computer. Even he was better than the driver who fell asleep.

Urmson and his colleagues learned that as their technology improved, the humans using it would become less reliable. "We realized that this was going to be much much harder to do than we had thought," said engineer Dave Ferguson. "It just became clear that, look, if we can't rely on the driver to pay attention, or to hand over control to, then basically we just have to solve the whole problem. We need to have this vehicle not need the driver."

Other factors weighed against the highway-assist feature idea. For one thing, the mainstream auto industry was developing similar technology. By this point, adaptive cruise control had been around for more than a decade, giving cars the ability to regulate their speed in relation to the vehicles around them. Now automakers including Mercedes-Benz, Audi, Cadillac, and Tesla were moving toward using camera systems to keep cars in their lanes, too. Even if Google's system performed better, it wouldn't look much more sophisticated. Plus, the idea of a highway-only feature annoyed Sergey Brin, who had greater ambitions in mind than making Google an auto industry supplier. If the tech giant was to be serious about driverless cars and splash into the auto industry, there was no point in starting with a halfway solution. Once more, Tony Tether's goal from the original Grand Challenge rang true: For this to be worth something, the vehicle had to be fully autonomous. Otherwise, what was the point?

While the Chauffeur team did its work, Google's leaders were exploring other ways to diversify their business. Since 2011, Sebastian Thrun had been serving as the company's first "Director of Other." Larry Page wanted more projects that didn't fit into Google's core, moneymaking online businesses. (Page was already dabbling in electric aviation for urban settings, a project that a few years later would be just one of many "flying car" ventures.) The result was Google X, a new division of the company dedicated to "moonshots," big bets on finding radical technological solutions for knotty, significant problems. Along with self-driving cars, X would work on providing internet to rural regions with giant balloons flying through the stratosphere, using drones to deliver packages and food, and making carbon-neutral fuel from seawater. After self-driving, its first high-profile project was Google Glass, a face-mounted computer that looked like a pair of eyeglasses. (Critics derided it as both creepy and buggy, deriding its users as "glassholes.")

Thrun led the Glass project for a while, but ended up in a more general role for the company. In 2014, he decided it was time to leave. Seven years after diverting from his academic career, Thrun was ready to go back to education, but not back to Stanford. His time at Google had left him with a taste for scale and an appreciation for unconventional approaches to making big things happen. The idea of working with a few graduate students at a time had lost its appeal. Instead, Thrun would run Udacity, an online education startup he launched to offer "nano degrees" in specialized subjects like data analysis, programming in Python, and developing Android apps. His role in the world of self-driving cars was done, at least for a while.

Thrun's fading from the scene left Chris Urmson in charge, and largely responsible for answering the question: What was self-driving good for? Making a highway driving feature was out. And while Larry Burns had gotten Chauffeur's engineers enthusiastic about the idea of pushing along an urban mobility revolution, the rigors of the effort so far had made clear just how hard it would be to make their vehicles work on a large scale, both efficiently and safely.

Urmson found a compromise. The idea came from Chauffeur's head of safety, Ron Medford. Medford had come to Google after serving as deputy chief of the National Highway Traffic Safety Administration, the agency that laid down the near-endless rules dictating how cars were built and certified, right down to the font on the vehicle identification number (sans serif). He told Urmson about a special category of low-speed vehicles, typified by the golf carts that senior citizens use to move around retirement communities. In exchange for a top speed of 25 mph, regulators exempted them from most safety standards, including crash testing. The low speed cap didn't seem like a problem; it would in fact mitigate the consequences of any crash. And the lower regulatory bar would make building a vehicle and putting it on the roads a whole lot easier.

His mind made up, Urmson called the team, now about seventy strong, in for a meeting. After years of modifying Prius and Lexus cars, they were going to design their own vehicle. One designed for self-driving. No steering wheel. No pedals. And they weren't going to sell it, either. They were going to operate just like Uber, but without the dude in the front seat.

The design team drew up an electric two-seat vehicle, roughly the shape of a child's Fisher Price car, but with doors, and painted white and gray instead of yellow and red. Google teamed up with Detroit-based auto industry supplier Roush to put it together. With the exception of the main Lidar—produced in-house by the team Anthony Levandowski had started at 510 Systems—which sat on the roof like a Shriner's fez, the sensors were tucked carefully into the vehicle's body. Urmson wanted it to "be approachable and friendly." There, it succeeded. The headlights, short-range Lidar, and grille formed a little face. Chauffeur called it Firefly. Based on its cuddliness and black "nose," the press (and in private, some Googlers) took to calling it the koala car.

Firefly's cutesy nature belied the fact that delivering on Urmson's lofty vision of a vehicle to usher in the age of autonomy had been no easy thing. For more than a year, the Googlers struggled with the vagaries of hardware development. Urmson wanted an exterior made of foam, so that any collisions, even at the vehicle's low speeds, would be less likely to hurt a pedestrian. It was a nice idea that didn't reckon with the difficulty of getting the right kind of foam, a material that would keep its shape over time but remain pliable enough to serve its purpose. Or the fact that when they did find the right balance, they couldn't get paint to stay on the exterior. Each iteration of the hardware design cost time and money. It was far from what you'd expect of the most sophisticated driving machine the world had ever seen. The interior felt cheap, with teal accents and big bulky buttons. In place of the steering wheel was a large bin where you might put some grocery bags, hardly an inspired choice from one of the planet's great technology purveyors. Ignoring the fact that Northern California gets hot during the day and cold at night, the design team didn't include a heating or air-conditioning system. Testing on chilly evenings, the safety operators would find themselves fogging up the glass, forcing them to open the windows and ride around in the cold.

And the bold statement of ditching the steering wheel was premature. For testing on public streets, the engineers still needed a way to manually direct the car. So they hacked in a flat disc that resembled the steering mechanism at Disneyland's Teacups ride. Spin it left to go left, and vice versa. Later, they added a joystick-like handle they called "the hot dog," necessitating more design iterations that cost more money and more time. The "cars" weren't very comfortable, and taught the software team little that they couldn't have learned with their existing test vehicles. The bigger problem that turned up in testing, though, was that 25 mph cap. The vehicles were allowed on roads with 35 mph speed limits—roads on which humans tend to drive at least 40 mph.

Before putting them on public roads, the Chauffeur team brought the Fireflies to the testing facility they'd opened in 2013, two hours southeast of San Francisco. Six years after his team had settled into the shuttered Castle Air Force Base in the run-up to the Urban Challenge, Urmson returned to the site. It had been a great testing ground then, and was even better once Google came in, cleaned away the masses of black widow spiders, and fixed up a few dilapidated buildings into proper offices.

The idea behind Castle was to have a complement to on-road testing and work in computer simulation. On the closed course, Chauffeur's engineers could run the world just as they liked, which was as wild as they could make it. As their cars ran around their ninety-one acres, negotiating roundabouts, merges, traffic lights, and crosswalks, workers jumped out of Porta Potties into the road, flung stacks of paper in front of the cars, and knocked piles of boxes into the street. Human-driven cars shot backwards out of driveways, cyclists swerved about, pedestrians jaywalked. All carefully planned drills, run over and over and over, to test the software's ability to handle them not just once, but every time. Steadily, the robo-cars improved, taking one step after another toward Urmson's goal of challenging Uber at its own game.

Firefly made its official debut on May 27, 2014, when Sergey Brin spoke on stage with journalists Kara Swisher and Walt Mossberg at Recode's inaugural Code Conference in Rancho Palos Verdes, California. After showing a video of Swisher riding in the vehicle, Brin, wearing a white T-shirt and black Crocs with a pair of Google's smart glasses hanging on his neck, talked up the advantages, like better sensor placement, of building a vehicle from scratch. He said Google planned to build one hundred to two hundred of the vehicles and to test them on public streets by the end of the year. Within a few years, he said, they might start running them without human backups. When they eventually went into service, they'd work as taxi-like vehicles.

Swisher asked the obvious question: "If you have the self-driving car fleet, don't you need to own the reservation system? Which is Uber. Which you have a big investment in." Brin punted, saying the team would sort out the details of the business closer to wide-scale deployment. "Longer term, it's not clear," he said. "We're almost certainly going to partner with a lot of companies, possibly Uber."

Travis Kalanick was at the same conference, and had received a heads-up that Google would be talking up a service that sounded a lot like a human-free competitor to Uber. He was furious, but when he took the stage, he said he loved the idea of plugging autonomous vehicles into Uber's network, and that it would be great news for riders, making Uber cheaper than owning a car. "The reason Uber could be expensive is because you're not just paying for the car, you're paying for the other dude in the car," he said. What he didn't say was that he wasn't going to wait to see the results of Brin's "possibly." If Google wasn't going to help him free up the driver's seat, he was ready to do it himself.

Carnegie Mellon University's Red Whittaker launched his career in robotics with a rover that went into the damaged nuclear reactor at Three Mile Island in Pennsylvania after a partial meltdown in 1979. *Courtesy of Carnegie Mellon University*

The members of Carnegie Mellon University's Red Team kept brutal hours in the run-up to the 2004 Grand Challenge, often working on their robot, Sandstorm, until they just couldn't anymore. *Courtesy of Carnegie Mellon University*

Anthony Levandowski got his self-riding motorcycle, Ghostrider, into the final of the 2004 DARPA Grand Challenge, but it flopped just feet from the starting line. *Courtesy of Jeff Rupp*

DARPA director Tony Tether had reason to celebrate Stanford's win at the 2005 Grand Challenge: The successful finish proved a major boost to DARPA's reputation and validated Tether's idea for an autonomous vehicle race. *Courtesy of the Defense Advanced Research Projects Agency (DARPA)*

Stanford's Sebastian Thrun was always glad to talk to reporters about Stanley, the autonomous vehicle that won the 2005 Grand Challenge. *Linda A. Cecero/ Stanford News Service*

The first iteration of Dave Hall's Velodyne spinning Lidar sensor weighed about three hundred pounds and collected 64,000 data points about its surroundings every second, in 360 degrees. Lidar soon became a must-have sensor for self-driving cars. *Courtesy of Dave Hall*

One hundred DARPA officials worked the 2007 Urban Challenge, many of them surveilling the course from the agency's mission control center. The agency spent more than $21 million putting the race together. *Courtesy of the Defense Advanced Research Projects Agency (DARPA)*

Anthony Levandowski, casually dressed for his television appearance, sent a self-driving Prius across the Bay Bridge as part of the Discovery Channel show *Prototype This! Courtesy of Joe Grand*

The early members of Google's Chauffeur team formed a murderer's row of computer science talent. *Left to right*: Chris Urmson, Jiajun Zhu, Dmitri Dolgov, Dirk Haehnel, Anthony Levandowski, and Nathaniel Fairfield. *Courtesy of David Goldwater*

After Chauffeur's Prius passed its first official driving test, Chris Urmson attached the special Nevada license plate anointing the car as a legal, road-going robot. *Courtesy of David Goldwater*

Using Lidar and other sensors, self-driving cars like Uber's identify the things they spot as cars, pedestrians, cyclists, or other categories. They navigate using pre-made maps, and their engineers gather detailed data on every move they make. *Courtesy of Uber*

Before Google's self-driving car team spun out as its own company, Waymo, the Chauffeur team sent a blind man named Steve Mahan for a carefully planned ride around Austin, Texas, in its Firefly vehicle, without an onboard safety operator to intervene if the car made a mistake. *Courtesy of Waymo*

General Motors had hoped to launch a self-driving service by the end of 2019 in San Francisco, using a version of its all-electric Chevy Bolt without a steering wheel or pedals. In July of that year, it pushed back the deadline indefinitely. *Courtesy of General Motors*

After Ford invested heavily in Bryan Salesky's startup, Argo AI, the two companies began self-driving tests on the streets of Pittsburgh, also home to Carnegie Mellon University. *Courtesy of Argo AI*

Argo remained an independent company outside the Ford hierarchy and far from Detroit—big factors in its ability to recruit software-savvy engineers. *Courtesy of Argo AI*

After leaving Google in January 2016, Anthony Levandowski created a robo-trucking startup called Otto and sent a big rig down a Nevada highway with nobody behind the wheel. *Courtesy of Uber*

In August 2019, Anthony Levandowski was indicted for allegedly stealing trade secrets from Google. Outside a San Francisco courthouse, he declared he was innocent. *Alex Davies*

— 9 —

Winner Take All

TRAVIS KALANICK RECOGNIZED THAT THE SELF-DRIVING CAR Google was creating was not just an opportunity for Uber. It was also a threat, and a potentially deadly one. Any ridehailing competitor that didn't have to pay humans to drive its customers would have a major cost advantage. If someone got there before Uber, they could do what Uber was doing to taxis: drop its prices to drive Kalanick's baby out of business. So in early 2014, Kalanick tasked Uber's chief product officer, Jeff Holden, with surveying the robotics world and scouting for a team that could rival the collection of DARPA Challenge veterans Sebastian Thrun had assembled. While there certainly was plenty of talent to be had—six teams had completed the Urban Challenge, after all—Holden soon homed in on Pittsburgh, the location of Carnegie Mellon University.

Uber made its initial move at the end of 2014, hiring the staff of a small Pittsburgh company called Carnegie Robotics. The company was run by John Bares, who'd started it in 2010, after spending most of his career at CMU. He'd worked for Red Whittaker as an undergraduate, helping to design and build the robots that went into the radioactive Three Mile Island nuclear site, then spent more than twenty years at the university. When Bares got an email from Uber, he didn't take it too seriously. Jeff Holden and his lieutenants said they wanted to build a self-driving car, but Bares and his colleagues replied that doing so would be far harder, costlier, and more time-consuming than Uber seemed to think. Holden persisted, making clear this was no lark: Uber would do whatever necessary to develop a self-driving car. Over a series of meetings that fall, the two sides came together. Once convinced of Holden's seriousness, Bares and his team got excited about the idea of using their expertise to improve the lives of everyday people. And Uber, with its existing fleets of drivers all over the world, offered a natural path to enter the market: As the robots mastered more kinds of roads and territories, they could, over many years, gradually take the place of those humans.

The Uber contingent acknowledged that chasing Google's effort—nearly six years old at this point—would take a serious investment. Kalanick was willing to make it, but he wouldn't be just one more customer of Carnegie Robotics. He wanted an in-house team, fully focused on creating the technology that would keep Uber relevant in the coming era. He hired Bares and most of Carnegie Robotics' staff, then licensed the company's intellectual property. But Bares knew they would need more firepower, and he knew where to find it. He started talking to his old colleagues at Carnegie Mellon University's National Robotics Engineering Center, which he had led for thirteen years before launching his own company.

Familiarly known as NREC (pronounced *en-reck*), the center had opened in 1996, housed in a large glass building on Pittsburgh's Allegheny River, a twenty-minute drive from the main CMU campus. It operated as a mostly autonomous arm of the university, in a distinctly nonacademic atmosphere. Its workers didn't do the kind of basic research that they could turn into dissertations. Success at NREC didn't mean landing a tenure-track position. It meant landing a huge contract and building a team that could deliver whatever that customer needed. A team didn't mean one professor and a handful of graduate students toiling away in a lab, but dozens of engineers taking what those labs produced and turning it into a commercial product that was durable, reliable, and affordable enough to convince a real customer to hand over real money. NREC had been the home of Bryan Salesky, Chris Urmson's reality-over-research lieutenant on CMU's Urban Challenge team.

During Bares's thirteen years as director, NREC worked for everyone from the army to John Deere to fruit farmers. For example, its strawberry plant sorter used machine learning techniques to recognize quality plants based on their size and health as they rode along a conveyer belt, then used air jets to sort them into piles. One team made a laser and GPS system that automated the process of counting trees, so orchard operators could keep inventory. Another developed magnetic robots that moved up and down the sides of warships, stripping their paint without damaging the steel. To get the paint off F-16 fighter jets, NREC made a laser-wielding robot. The lab built autonomous forklifts and mining vehicles. Its Pipeline Explorer roved through high-pressure natural gas lines, looking for trouble. SmartCube and CognoCube monitored animals involved in drug testing trials.

The NREC bot you'd least want to meet in a dark alley was Crusher: a six-wheeled, hybrid-powered, and fully autonomous military beast with a unique suspension that let it smash through trenches and over rock piles. True to its name, it had an uncanny ability to steamroll over cars. NREC created it for—who else—DARPA. In an era when venture capitalists had little interest in robotics, NREC was a vital force for putting new technology into the market.

Nowhere else, Bares knew, would he find so many specialists not just in making damn good robots, but in making them work commercially. As 2014 drew to a close, Bares and Jeff Holden started talking to NREC's workers about Uber's new venture. It wasn't a hard sales pitch. The NREC people would get to keep working on robotics, with a singular focus on autonomous driving on civilian streets. Uber, having raised another $1.2 billion in funding that summer and desperate to catch up to Google, would double, maybe triple their salaries. And they wouldn't have to move, or even adjust their commutes. San Francisco–based

Uber would open a new engineering center next door to the NREC lab, in an old chocolate factory.

Over the years, NREC had seen plenty of its employees leave for other jobs. It had never seen dozens leave for the same gig, all at once. In February 2015, about forty NREC employees, including the lab's director, resigned. Together, under John Bares's leadership, they formed a new arm of Uber, the Advanced Technologies Center. Nearly thirty years after the blue Chevy panel van known as NavLab 1 explored Carnegie Mellon's campus at an octogenarian's walking pace, they were going to bring self-driving cars to Pittsburgh.

The news of Uber launching an outright competing effort hit the Chauffeur team like a cannonball to the stomach. On paper, Chauffeur was rolling. It had hundreds of cars and dozens of engineers. They were on the verge of clocking their 1 millionth mile on public roads, the equivalent of what an American driver covers in seventy-five years. Their cars were averaging more than three thousand miles between times when the safety operators disengaged for a safety reason. In March of 2015, Chris Urmson gave a TED talk about Chauffeur. He revealed the 2013 highway trials with the distracted Googlers and explained why his team was focused on taking the human entirely out of the driving process. He talked about the 1.2 million people around the world who die on the road every year, the 6 billion collective minutes Americans spend commuting every day.

On stage in Vancouver and displaying a digital feed of scenes the cars had encountered, Urmson rattled through examples of just how crazy driving in the human world really was: "Watch to the right as someone jumps out of this truck at us. And now, watch the left as the car with the green box decides he needs to make a right turn at the last possible moment. Here, as we make a lane change, the car to our left decides it wants to as well. And here, we watch a car blow through a red light, and yield to it. And similarly, here, a cyclist blowing through that light as well." One day in Mountain View, the car had to stop for something it had never seen before: a woman in the street in an electric wheelchair, chasing a duck in circles.

Urmson closed his talk with a photo of his two boys, both of them born while he was helping lead Carnegie Mellon's Red Team through the Mojave Desert. "My oldest son is eleven, and that means in four and a half years, he's going to be able to get his driver's license," he told the audience. "My team and I are committed to making sure that doesn't happen."

Offstage, Urmson's view of the project wasn't so rosy. Chauffeur's software was improving, but the effort was slowing down just as Uber's was ramping up. Uber was pulling in self-driving talent like a tractor beam; it would soon have three hundred people. Bares's team had established a fifty-four-acre test track, hired a crew of safety operators to test its cars, and retrofitted a fleet of Ford Fusions with all the necessary sensors, including Dave Hall's Velodyne Lidar laser scanners.

Six weeks before his TED talk, as Uber revealed it was turning the push for self-driving tech into a race, Urmson sent an email to Larry Page, Sergey Brin, Google financial chief Patrick Pichette, and Google X lead Astro Teller. "Over the last few months we have begun operating to minimize downside, not to win," he wrote. "We risk wasting our advantage and investment if we do not empower the team to move with velocity." Uber was hiring people he had wanted to bring on board a year and a half earlier but couldn't because of hiring freezes. The team's morale was being damaged by budget constraints, he said.

Chauffeur was confronting a nasty truth: In developing autonomous driving machines, the easy work came first. A small but talented team could make a car navigate any road Larry Page picked off a map. But turning that achievement into a commercially viable product—a car that handled every road, every situation, every time—would require years of work, billions of dollars, and hundreds if not thousands of workers. It required chasing down an endless series of what engineers call "edge cases." Here, that meant all the strange things that happen on the road.

Oftentimes, some element of the system would plateau, reaching a point where the engineers just didn't know how to make it any better. The car might be able to identify a motorcyclist 999 times out of 1,000, but that wasn't good enough to be a safe driver. Getting to 9,999 times out of 10,000—closer to acceptable—might require the use of a new technique that the software wasn't designed to include. Adding it might mean tossing out code the team had written over months or even years, and rebuilding it from scratch. That was a hard decision to make, especially since ripping up any element of a driving system could slow the development of all the bits that interacted with it.

For all their work, the paradox laid bare by Hans Moravec decades earlier still rang true. The things that humans do so easily, adapting to novel circumstances and moving through them, all too often brought robots to their knees. And now the Chauffeur team had to fight not just the problem itself, but a competitor flush with talent and ambition.

In his email, Urmson broke from his usually subdued demeanor and channeled Red Whittaker's bombast, pushing Google's leadership to commit to finishing what they had started. "We have a choice," he wrote, "between being the headline or the footnote in history's book on the next revolution in transportation. Let's make the right choice."

Anthony Levandowski, too, worried that his team was squandering its head start. No one doubted Urmson's engineering skills. But some team members, including Levandowski, found their leader overly focused on research, too invested in the idea of perfection, to the detriment of making a product they could sell. It was a tendency that traced back to the Urban Challenge, when Urmson and Bryan Salesky argued over the balance between advancing the technology and delivering a race-ready vehicle. Now, without a DARPA-imposed deadline, Levandowski and others felt that Chauffeur had become an "eternal research project," abetted by Google's long leash and ample resources. Mocking the project's penchant for burning through financial reserves, members of

Chauffeur's electrical engineering team—led by Andrew Schultz, who had cofounded 510 Systems with Levandowski—took to wearing shirts that read "Money Destruction Team," depicting piles of flaming cash.

At some point, Levandowski thought, the team had to take concrete steps toward introducing a commercial product. But he had little power to wield. Despite years of machinations—or because of them—he found himself sidelined. He had lost his role as hardware lead to Salesky in early 2013. Salesky had first joined the team in 2011, but quit a year later, telling Larry Burns he was tired of dealing with the tension between Urmson and Levandowski. After Thrun, who had championed Levandowski, left the project, Urmson wooed Salesky back to the team, putting him in charge of hardware. Levandowski was demoted to manager of Chauffeur's efforts to design and develop its own Lidar laser scanners, building off work begun at 510 Systems. Not much for hierarchies, he made a habit of going over others' heads, emailing Larry Page his thoughts about the project and ideas for moving it forward.

Frustrated by the team's lack of progress and his diminished role, Levandowski began considering other options. He had talked about leaving Chauffeur before, making plans to start new efforts to compete with the effort he'd helped launch. The difference in 2015 was that the choice was no longer between staying with Google and going it alone. Uber was in the mix now, with a mountain of money to spend and a CEO who believed that delivering a self-driving car was crucial to his company's future. Travis Kalanick had hired dozens of Carnegie Mellon researchers to get things started, and remained willing to talk to anyone who could help him eat into Google's head start. Especially anyone who had helped Google build that gap to begin with. Levandowski started meeting with Uber executives, including Jeff Holden. Pierre-Yves Droz, the 510 Systems cofounder leading Chauffeur's Lidar work, later said that Levandowski told him Uber would be interested in "buying" their Lidar team.

Urmson said that after he took the top spot, he tried to make peace with Levandowski. "I worked very hard to kind of mend bridges and bring him, you know, into the fold." But when he found out about Levandowski's latest maneuverings, he was fed up. "We need to fire Anthony Levandowski," Urmson told some colleagues in an August 4 email that was later introduced in court. "I have just heard today from two different sources that Anthony is approaching members of their team attempting to set up a package deal of people that he could sell en masse to Uber." Larry Page didn't receive that email, but he was used to chatter about Levandowski's character and ethics. "There was always concern," the Google CEO said. "You know, people would say, 'Oh, Anthony is doing this. Anthony is doing that.'" But the company leadership thought Levandowski—who had a long track record of getting things done—was worth keeping in house.

The next month, Google's higher-ups made a different kind of staffing change at Chauffeur. They brought in a new boss. John Krafcik may have studied at Stanford, but he was a tried-and-true car guy. He'd started his career at the New United Motor Manufacturing plant that Toyota and General Motors ran together in the 1980s, near San Jose. Krafcik then got a master's in management at MIT, where he created the term "lean production" to describe Toyota's mass manufacturing mastery. He spent the 1990s at Ford, running the development of the Expedition and Lincoln Navigator SUVs—crucial profit centers for the automaker. From there, Krafcik moved to Hyundai, where he became CEO of the Korean automaker's American operation in 2008. The end of his tenure there was marred by the revelation that Hyundai and its sister company Kia had overestimated the fuel economy of nearly 1 million cars sold between 2010 and 2013. The companies blamed the discrepancy on "procedural errors," and while Hyundai said the scandal had nothing to do with it, it declined to renew Krafcik's contract at the end of 2013. Krafcik ended up at True Car, which collected sales data from dealerships to help car buyers find good deals. It wasn't the greatest result for an exec who'd been named among potential successors to Ford CEO Alan Mulally. But Krafcik looked to have the relevant experience that the Google team would need to bring a car to market, such as knowing how to negotiate the industry's intricate regulations and supply chains. His hiring was a clear indication that Page and Sergey Brin were eager for Chauffeur to start producing revenue, and that they didn't believe their software engineers had the business chops to make that happen. Krafcik was named the project's first CEO, effectively demoting Chris Urmson to the technical lead.

After the struggle to design the Firefly car, Krafcik focused his early energy on finding Chauffeur a partner to help produce vehicles for a self-driving service, and negotiating a potential deal with Ford. But the personnel issues that preceded his start date continued to be a problem. In October, Google X chief Astro Teller emailed Krafcik and Urmson to say that Larry Page was concerned about Levandowski leaving and going to help the competition. Around this time, Page said, Levandowski approached him about his desire to leave the team and go start his own venture, working on autonomous semitrucks. He was tired of his colleagues, Page recalled him saying, and said something to the tune of "Why don't I just go to a company that does trucking? And everything will be fine." Page told him that would not be fine—that it would be a clearly competitive threat. "I now believe he was trying to get me to say that it would not be competitive if he did trucking, because he was already doing it," Page said in a 2017 deposition.

It was a reasonable theory. Over the summer and fall, Levandowski's discussions with Uber's executives escalated from talking about collaborations between the ridehail company and whatever effort Levandowski launched, to having Uber acquire it outright. Levandowski kept talking to his Google colleagues about having them join him when he left, though he didn't say the plan was to sell something to Uber, and later insisted he was never committed to working with Kalanick. In December, just before the holidays, between fifteen and twenty people gathered at Levandowski's house for a paella dinner. At least half a dozen of them worked for Google, and were interested in an effort that would accelerate their technology's move into the commercial world. There, Lior Ron, a Google Maps executive whom Levandowski knew from his work on Street View, gave a presentation about automated trucking that fit the bill nicely.

If the goal was to commercialize self-driving technology, trucks presented a tantalizing opportunity, especially compared to a taxi system. The 18-wheelers spent nearly all their time on the highway, where driving was simplest. And the economic imperative was clear. Trucks moved more than 70 percent of freight in the US, bringing in $700 billion a year in revenue. Trucking companies gladly shelled out extra cash for cruise control and advanced safety features, anything to make their drivers safer and reduce the hefty costs that came with any crash. Yet there weren't nearly enough drivers willing to move it all—the American Trucking Associations trade group reported that in 2015, the industry needed forty-five thousand more drivers, and that the shortage would only worsen with time. Some years, the turnover rate for long-haul drivers topped 90 percent. To accommodate a robotic solution, trucks would require slightly different sensors and driving styles (to see farther and start braking earlier, for example) than cars, but the fundamentals were the same. And the route to success seemed a whole lot clearer than what Google was doing, as it moved away from the Firefly vehicles and toward a deal with some automaker.

As Levandowski recruited people to his latest venture, his talks with Uber escalated to conversations with CEO Travis Kalanick. The engineer and the executive had met earlier, at Sebastian Thrun's suggestion, and hit it off. They would meet at San Francisco's Ferry Building and take long walks along the waterfront, toward the Golden Gate Bridge. They exchanged hundreds of text messages and spent "jam sessions" brainstorming. Levandowski saw Kalanick as something of a mentor; Kalanick called the younger engineer his "brother from another mother." Each was aggressive in his business dealings and willing to push up against the limits of what others considered ethical, if not legal. They saw self-driving as it was during the 2004 DARPA Grand

Challenge: First place got the prize, with nothing left over for anyone who pulled in second. "Autonomous transportation is very possibly a winner-take-all and, thus, existential for Uber," Kalanick told his team. And Uber wasn't looking like a winner. A year in, Holden told Kalanick in an email, "We started from a huge gap with Google, and I think we've all been sobered by how hard it is to close that gap, even with exceptional effort." In his notes, team leader John Bares wrote, "My strain, personal strain, increasing pressure to catch up seven years and deploy 100,000 cars in 2020."

For Kalanick, bringing Levandowski aboard was an obvious way to supercharge his team's bid to be that winner. He considered the Googler one of the world's premier experts on the subject, someone who had been studying the problem for more than a decade and knew how to get an operation off the ground, fast. Levandowski's expertise was particularly important with regard to Lidar, the tool that he had spotted back in 2005, had helped sell to all the top Urban Challenge teams, and had developed for himself at 510 Systems before bringing it to Google. The tool that, more than any other, Chauffeur's cars relied on to move through a crowded, chaotic world. On the first Sunday of 2016, Levandowski went to Uber's office for another jam session with Kalanick. On a whiteboard, Kalanick wrote: "Laser is the sauce."

Kalanick's do-or-die attitude appealed to Levandowski. "Uber had to ship a product," he said. This would be no eternal research project. And it could pay handsomely. Back at the whiteboard, under "possible outcomes," Kalanick wrote: "Uber Super Duper, or USD, so could be Code Name $." The idea was straightforward: Levandowski would leave Google and start his own company, which Uber would then acquire.

On January 7, Levandowski emailed Larry Page to wish him a happy new year, and to offer what seemed, in retrospect, a final olive branch. "Chauffeur is broken," he wrote. Some team members were afraid to put their tech on the market. "We're losing our tech advantage fast." He had put these concerns to John Krafcik, he said, but the Chauffeur boss was preoccupied with a potential deal with Ford. Levandowski floated the idea of starting a "Team Mac" within Chauffeur. Page forwarded the email to Krafcik, who as a Silicon Valley newcomer didn't understand the bit of technology lore that Levandowski was referring to: In the early 1980s, Apple was working on a new PC to be called the Lisa, but development was slow, the price tag was approaching $10,000, and IBM was dominating the personal computer market. Steve Jobs wanted to make a much cheaper and superior product. While the main team kept going with the Lisa—which turned out to be a flop—he put a few Apple employees to work on what became the Macintosh, one of the most successful computers of all time. Levandowski thought he could do something similar within Google, just not within Chauffeur. The idea went nowhere.

On January 27, Travis Kalanick received an email from an exec on Uber's deal team: "We have a tentative deal with respect to Newco," using their placeholder name for the company Levandowski would launch. That same day, Levandowski emailed Page again, saying "there's just too much BS," with Urmson, Krafcik, and Salesky. He had bumped into Sebastian Thrun, he said, who seemed very happy working on his education startup, Udacity, as well as a new aviation project for Page, a flying car venture called Kitty Hawk. Maybe it was time for him to do something new as well.

Levandowski had started work on driverless vehicles with a wild idea for a motorcycle all the way back in 2003, convincing enough people to share their time and resources to make it real. He had helped Dave Hall sell his groundbreaking Lidar sensors and helped Google map the world. He had sent an autonomous car over the Bay Bridge in 2008, showing the world that the image put forward by the DARPA Challenges wasn't dead. And he had spent nearly a decade with Google, trying to deliver on the dream, arguing and scheming and fighting for a brazen endeavor, never flinching at difficulty. After doing all that, he had lost the fight for command and been pushed aside. But he'd never been one to "vest in peace"—Silicon Valley–speak for sitting quietly and watching the money pile up.

It had been four years since the brokering of the Chauffeur equity deal. The end of 2015 was bonus time, and Google, sticking to the instructions Page had issued in 2011, made Anthony Levandowski rich. They pinned the value of the self-driving effort at $4.5 billion, which translated to $120 million for the engineer. With the money finally coming his way, the motorcycle-tuning, sensor-selling, Lidar-creating, business-begating Levandowski quit. "I want to be in the driver seat, not the passenger seat, and right now [it] feels like I'm in the trunk," Levandowski emailed Page. He was striking out on his own, he said, with a self-driving truck outfit. When Chris Urmson heard the news, he had Levandowski gather his things, and escorted him out of the office.

What Urmson didn't know was that more than a month before Levandowski resigned, Waymo later alleged, he had downloaded a program called TortoiseSVN, software that granted access to Chauffeur's file server. At 6:40 on a Friday night, he connected to that server and downloaded 14,107 technical files, some of which documented the Lidar systems his team had spent years developing. He connected an external hard drive card to his laptop the following Monday morning, and unplugged it nine hours later. A few weeks later, the day after the "laser is the sauce" meeting with Kalanick, Levandowski exported a few more files from a shared Chauffeur drive, whose titles included terms like "Intensity Calibration," "Extrinsic Calibration," and "tuning instructions." And in mid-January, two weeks before Levandowski resigned, he exported a file called "Chauffeur TL weekly updates—Q4 2014," which detailed the latest work of the team's technical leaders.

On February 1, Levandowski officially formed Otto Trucking with Google mapping exec Lior Ron, Google roboticist Claire Delaunay, and Chauffeur software engineer Don Burnette, who had long advocated for a more product-minded push toward commercialization. According to Burnette, when he resigned from Google, a few weeks after Levandowski's departure, John Krafcik asked him to lunch and warned him against starting a rival effort. Google was going to own the autonomous driving space, Krafcik said. Whatever effort Burnette launched or joined, Krafcik told him, Google would "crush" him.

In May, Levandowski and Ron revealed the company to the public. Of its roughly thirty employees, half had come from Google. "We're a team comprised of the sharpest minds in self-driving technology, and we are committed to reimagining transportation—not just improving it," they wrote in a blog post. "We are at Otto because we're driven by an urgency and deep obligation to accelerate the future. . . . It's time to move." The post included a video of an 18-wheeler with an array of sensors on the cab rolling down Interstate 80 near Reno. It showed a man doing paperwork in the back seat of the cab, with nobody behind the wheel.

The demonstration's legality was dubious: Otto hadn't taken the time to apply for a permit to test its autonomous vehicle in Nevada or passed the kind of driver's exam Google went through in 2012. It didn't have the special red license plate with the infinity symbol. The DMV officer in charge of the autonomous testing program was furious. Otto argued that because someone was in the cab, it didn't count as an autonomous vehicle. The point was moot though, because Nevada's law didn't include any penalties for violating its rules.

By the spring of 2016, Otto had a fleet of six trucks plying the highways of the Bay Area, while the software engineers, operating out of offices in San Francisco and Palo Alto, pushed weekly and monthly updates. But Levandowski & Co. waited until mid-August to announce their big news: They were being acquired by Uber. After the tentative "Newco" deal in January, Otto and Uber had made things official in April, but kept quiet. At 1 percent of Uber's massive valuation, the deal was worth roughly $680 million—an astonishing figure. But, as at Chauffeur, it hinged on hitting goals. Levandowski and his band wouldn't see major dollar signs until they delivered the top-flight Lidar laser scanner that Kalanick knew Uber needed.

John Bares knew Levandowski a bit. He'd attended the Grand Challenges, and when he saw the young engineer who had made a self-driving motorcycle, he'd encouraged him to move to Pittsburgh and take a job at NREC. Now Levandowski would be Bares's boss, and when he visited the Pittsburgh team, Levandowski wasn't impressed. "Wow," he texted Kalanick in May. "I am super pissed at what is going on at ATC. There's one one [sic] who is pushing for the right things." It was a harsh but not unreasonable assessment. The former Carnegie Mellon researchers were terrific roboticists, but hadn't put a real dent in Google's lead. They had barely done any testing on public roads. And they certainly weren't used to working with the kind of speed and urgency that Levandowski brought to his efforts.

In August, Uber announced more than the Otto acquisition. It was opening a research lab in San Francisco to work with the one in Pittsburgh. And within a few weeks, Uber would launch its self-driving Ford Fusion sedans and Volvo XC90 SUVs on the streets of Pittsburgh. They'd be part of Uber's regular service, with safety operators up front and free rides for passengers who happened to order one. Uber, not Google, would be the first company welcoming the public into its autonomous vehicles in the United States. And the man in charge of the whole effort was Anthony Levandowski.

Taken together, the news was a major coup for Uber, dealing a blow to the perception of Google as the undisputed leader in self-driving tech. It rattled the Chauffeur team, too, and came at a bad time. Levandowski wasn't the only core team member to leave Chauffeur after bonuses were doled out. And he wasn't alone in thinking a nimbler, more targeted effort could make progress where Google had struggled. In August, longtime engineers Dave Ferguson, Jiajun Zhu, and Russ Smith quit to start their own venture focused on what they considered a near-term achievable commercial enterprise: local deliveries using low-speed autonomous vehicles. Salesky, the hardware lead, followed them out the door a few weeks later, intent on doing something where he'd have more say and wouldn't have to figure out how to build his own car.

They had plenty of places to go. Google might not have delivered a self-driving car yet, but its efforts had woken the world to the technology's potential. Venture capitalists were looking to fund new startups, and major corporations that had dismissed the idea of robotic cars were planning to catch up.

The most significant departure that summer was that of Chris Urmson. The longtime leader had lost his top spot to Krafcik, and by some accounts the two didn't see eye to eye. Either way, after seven and a half years with Google, Urmson wanted a change. "After leading our cars through the human equivalent of 150 years of driving and helping our project make the leap from pure research to developing a product that we hope someday anyone will be able to use, I am ready for a fresh challenge," he wrote in a blog post announcing his departure. "If I can find another project that turns into an obsession and becomes something more, I will consider myself twice lucky." Dmitri Dolgov took his place as technical lead.

With the news that Uber had brought in Levandowski and was opening a research center in the Bay Area, the remaining Googlers worried about losing their potency. The rest of the world was catching on to the potential of self-driving technology, and Google was no longer the only game in town. "For the more senior folks on the team, the recent departures of Dave, Jiajun, Russ, and Chris have really brought into focus the path of starting something new with real ownership (or joining forces with the people who already left)," Dolgov told Krafcik and others in an August email. "It's clear how active the space is right now with VCs & large companies waving crazy opportunities around." Uber was actively looking to poach Google engineers, and the pitch sounded better now that it didn't involve moving to Pittsburgh. Chauffeur, which had temporarily frozen hiring, lost more than a few people to its rival. "I think we should be seriously concerned," Dolgov wrote. Another factor he didn't mention was Google's generosity. The team's core members had all made millions of dollars from the lucrative bonus program, what staffers described as "F-you money." Enough to alleviate concerns about job security and open them to new, riskier ventures where they could run things the way they saw fit.

Now in charge of Uber's self-driving effort, Levandowski wanted to bring the robots not just to Pittsburgh, but to his new employer's hometown—and to Google's backyard. On December 14, 2016, Uber announced that it would add sixteen self-driving Volvo SUVs to its fleet of cars in San Francisco. As in Pittsburgh, the cars would have safety operators behind the wheel, and passengers who happened to call one would ride for free. It didn't take long for trouble to hit. California DMV's chief counsel deemed the operation illegal, because Uber didn't have a permit to test self-driving tech in the state. It had been running cars in San Francisco for weeks, and hadn't even bothered to apply for one. The DMV demanded Uber stop its tests, threatening legal action.

Levandowski cared little for regulation. "I think there should be no rules," he once said. And he didn't back down. California's law defined an autonomous vehicle as "any vehicle equipped with technology that has the capability of operating or driving the vehicle without the active physical control or monitoring of a natural person." Because Uber's cars weren't sophisticated enough to drive without a human backup, Levandowski argued they didn't fit that definition, and thus didn't need the permit. But San Francisco mayor Ed Lee, widely regarded as a friend to big tech, called on Uber to stop testing. California attorney general Kamala Harris's office threatened to seek an injunction against the company. Uber stood its ground, but the popular support it had often relied on in fights with local regulators took a hit when one of its cars blew through a red light the same day it announced it was welcoming riders aboard. The moment was recorded by a local taxi driver, who gladly shared it with the media.

Levandowski's loophole-based argument might have stood up in court, but he lost the fight when the DMV played its trump card a few days later, revoking the registrations for the sixteen test cars. After the weeklong tussle, Arizona governor Doug Ducey publicly invited Uber to come next door. Ducey had issued an executive order requiring various agencies to do whatever necessary to allow for testing in the state. In Arizona, Uber didn't need to pass any special tests or report any data. Levandowski had his Volvos loaded onto an Otto-branded truck and sent them east—letting the 18-wheeler do the driving.

A Long, Slow Donkey
Ride Through Hell

GOOGLE'S MAY 2014 REVEAL OF FIREFLY, THE KOALA-LIKE VEHICLE without a steering wheel or pedals, had provoked Travis Kalanick into creating a self-driving effort to rival Chauffeur, based on the idea that if you can't join 'em, beat 'em. It got reporters speculating about the logistics of how Google would run a robotic ridehail service. It sparked critiques from designers. And it caught the attention of auto industry analyst John Casesa.

Over the past few years, Larry Burns's vision of cities shifting from personal cars to fleets of small, electric, shared, and autonomous vehicles had started to look like an impending reality. The economic fallout from the Great Recession tempered young people's ability to buy cars. The same Millennials left the suburbs for the cities, where Uber, Lyft, Zipcar, and other services made car ownership optional anyway. Casesa had caught on and started talking to his clients about a future where selling more and more cars wouldn't be enough to stay relevant, or even solvent. But like Chris Urmson and Anthony Levandowski visiting Detroit in those early days, he found his soothsaying either dismissed or ignored.

In Firefly, Casesa found tangible proof that the transformation was real, and imminent. Google was serious about this self-driving business, and after five years of work—the time it took an automaker to develop a conventional car—it had the product that would bring it to market. The business Chauffeur proposed threatened not just to disrupt the way Detroit operated, but to obliterate it. "The self-driving car wasn't an enhancement of the existing technology," Casesa said, the way cruise control had built on what automakers already had. "It was a substitute for it. It was like going from the horse to the car."

One of Casesa's big clients, the Ford Motor Company, wasn't in much of a position to adapt to this new threat. In the mid-2000s, it had barely been staying afloat in its native business model. Lackluster products produced lackluster sales. Executive infighting hampered various attempts at change. Generous union contracts put Ford at a disadvantage to automakers like Toyota, who built their cars in factories whose workers were rarely unionized. Between 2001 and 2006, Ford's stock price fell 65 percent. The company that had first sold America on the idea of putting a car—or three—in every driveway was on the road to bankruptcy.

In 2006, CEO William Clay Ford Jr. realized he wasn't the man to save the company his great-grandfather Henry had built. He convinced Alan Mulally, a longtime Boeing executive, to take the helm. One of Mulally's first tasks was to walk into the ballroom of New York's Marriott Marquis, stand up in front of more than four hundred bankers, and ask for a loan. In exchange for $23.6 billion, Ford posted as collateral just about everything it owned. All its domestic assets. Its factories. Its patents. Even the trademark to its blue oval logo. It was an act of desperation, an all-in bet on Mulally's plan for rebuilding Ford into a semblance of its former self.

It worked. Mulally focused on core Ford products, selling off Volvo, Aston Martin, Jaguar, and Land Rover—brands acquired in a hazy-eyed attempt at conglomeration. He closed factories and rewrote deals with workers. He built a new, united executive team. By 2009, the automaker was turning a profit, even as the financial crisis induced a drop in car sales. Where General Motors and Chrysler went bankrupt, Ford stayed solvent. When Mulally stepped down in July of 2014, he was lauded as a hero for pulling a giant off its deathbed and saving who knew how many jobs. But his healing efforts had focused on making Ford as good as it had once been at building cars and trucks. While Ford was anesthetized and under Mulally's knife, notions of transportation had evolved. It woke up in an age where being an old-fashioned automaker might not be enough anymore.

Others in the auto industry's intelligentsia saw what Casesa saw. Analysts at KPMG and Morgan Stanley warned that the era of the two-car family was on its way out, and advised automakers not to rely on the rising middle classes in China and India for new customers, especially not in urban areas—cities like New Delhi were already too choked with emissions and traffic to allow a new influx of personal vehicles. But the car business hadn't fundamentally changed in one hundred years. Automakers made a small variety of makes and models. They occasionally introduced new safety, comfort, and tech features that, over years, moved from luxury to down-market rides. They ran endless television commercials showing their sedans and SUVs cruising through mountains and along coastlines, playing on the oh-so-American insistence on personal freedom. And when someone did leave the dealer lot with a new set of keys and a loan, the company hoped they would come back a few years later for another one. "The auto industry is a century-old ecosystem being ogled by outside players hungry for a slice of a $10 trillion mobility market," one analyst wrote. "Many want in. It's just beginning. And it won't stop."

Many did want in, and the door had swung open. Venture capitalists started hearing from startups focused on self-driving cars, and cutting checks. Premier among these newcomers was Zoox. Led by Jesse Levinson, a former protégé of Sebastian Thrun's and a member of Stanford's 2007 Urban Challenge team, it proposed not just teaching cars to drive, but redesigning the very shape of the car (without saying much about what that meant). After raising $40 million in 2015, it pulled in another $250 million the following year, then another $500 million in

2018. Even smaller efforts like nuTonomy, founded by a pair of MIT roboticists, or Drive.ai, run by Stanford deep-learning experts, pulled in millions to compete with Google and Uber.

Casesa got his chance to better help Ford compete in 2015, when Mark Fields, who succeeded Mulally as CEO, hired him to be the company's first head of global strategy. Casesa's charge, as the public relations team put it in a press release, was to "enhance existing business strategies and to identify and evaluate new opportunities leading to profitable growth." Opportunities like getting Ford into the brave new world of self-driving cars. Ford was no stranger to autonomous technology. Its researchers had participated in the 2005 Grand Challenge and 2007 Urban Challenge, making it to the final round of each. The team came away having learned a lot about how to make a car drive itself, and having realized just how much more fundamental work needed to be done before the technology would be viable. After the 2007 race, they started chipping away at those basic problems, but full autonomy wasn't the goal. Their bosses were mostly interested in what limited features they could extract from the greater project and put in their cars to pad the bottom line. They kept working with the same F-250 they'd run in the Challenges, fiddling with its ability to map its surroundings and determine its own position within them. They worked on the best ways to calibrate and extract data from their Lidar, camera, and radar sensors, all the stuff they needed to get right before autonomous driving could work.

Those years of fundamental work were necessary because for all the Urban Challenge had shown to be possible, it had masked the gulf between a demonstration and a viable product. As a onetime event in specific, constrained conditions, the race let its contestants elide some of the work that came with making a robust, reliable system. The fact that six teams finished the course was remarkable, but the informed observers knew that didn't mean the problem was anywhere near solved. "DARPA was fanning the flames really hard. They were on the bellows, cranking away," said Ed Olson, a roboticist who worked on MIT's Urban

Challenge team, then with Ford while at the University of Michigan. When DARPA stopped cranking, the people who wanted to keep pushing the technology took stock of what a further push would require. "We had to pay down a lot of technical debt," Olson said. Ford, though, wasn't overly invested in making those payments. The research budget had a lot of areas to cover: engines, catalytic converters, fuel efficiency, recyclable materials, paint, and so on. The automaker treated the shift away from human driving as a less than vital research problem, worthy only of the part-time attention of a few in-house engineers.

The shifting transportation landscape affected more than whether a human or a robot did the driving, and Ford's new CEO realized as much. With swept back hair and a closet full of fine suits, Mark Fields was the classic image of a Detroit car exec. But in January of 2015, he flew to Las Vegas to give the keynote talk at CES, the massive consumer electronics showcase. This was news that wouldn't wait until the Detroit Auto Show a week later. Fields positioned Ford as hip to a changing world, and no longer content to be a mere automaker. "We're driving to be both a product and a mobility company and, ultimately, to help change the way the world moves," he told the crowd, filled with tech journalists.

Later that month, Fields went to Palo Alto for the grand opening of Ford's Silicon Valley Research and Innovation Center. Before then, the site had consisted of a team about a dozen strong, working in an office above a Wells Fargo bank. Fields had visited during the first few months of his tenure as CEO, and had now expanded the outpost into a proper base with a garage and plans to hire more than a hundred people, some of whom would work on the company's modest existing autonomous driving research. Many of them would be young software engineers, far easier to hire in the land of tech than in Detroit. Ford was no pioneer here. Volkswagen, BMW, Nissan, Audi, Toyota, Mercedes-Benz, and General Motors had all established outposts in the Bay Area years earlier, eager to mine that talent and to connect with the tech companies whose products could make their cars more compelling.

They all needed the help. The men and women of the auto industry were remarkable when it came to turning tens of thousands of parts into a vehicle that could drive at highway speeds and protect its occupants in violent crashes. But when it came to software, they weren't on the same level as Silicon Valley. The flagship example was the "infotainment" systems that ran on cars' center screens, with features like navigation and connection to one's phone. Automakers struggled to make systems that were capable and easy to use, frustrating customers who'd paid extra for them. Like everything in the auto industry, the systems spent years in development, so whatever made it into a car was far from the latest and greatest tech. The difference between the screen in a driver's hand and the one in his car was glaring.

That difference took on new importance in 2014, when Google and Apple introduced their automobile entertainment alternatives. Android Auto and Apple CarPlay let drivers project their phones onto their car's center screens, giving them access to maps, phone contacts, messages, music, and other apps. They brought to the car operating systems that users already knew and liked, and whose quality would keep up with their phone's. The popular response was summed up in the title of a report by research firm Strategy Analytics: "Consumer Interest for In-Car Smartphone Mirroring Is Almost Universal." What scared the automakers wasn't just that these options sapped customer interest in paying extra for their own systems, but that they ceded control of a major part of the in-car experience to Google and Apple. In the age of the smartphone, when drivers spent far more time in traffic—a full week every year, by some accounts—than on winding backroads, tech like infotainment mattered more than horsepower. That was a problem for companies that had spent decades and millions of dollars establishing brand identities with slogans like "Go Further" and "Find New Roads." And it was just a preview of what might happen next.

Uber and Lyft riders didn't care whether they were riding in a Toyota Prius or a Honda Fit. To them, the car was an interchangeable commodity. The important variables were the price and quality of the ride. If Google delivered on self-driving (a technology Apple was also developing, more quietly and less successfully), that could be the future of all cars. Players like Ford would be reduced to suppliers, doing the hardware grunt work while the tech companies made what mattered, and what paid. That was bad for both business and the egos of the people running the manufacturing giants that in the twentieth century had been emblems of American might and prosperity. Going in on "mobility" was an opportunity to flip the script. But what that meant, exactly, was hard to suss out.

When Fields declared Ford a mobility company as well as an automaker, he announced that the company was running twenty-five "experiments" as an early step in figuring out this new duality. It was operating remote-controlled cars with 4G networks, letting drivers program their thermostats from the dashboard, improving voice recognition software, developing tech to identify open spaces in parking lots, running a car swapping program among employees, and more. These ideas were intriguing, and they used all the right buzzwords: data, experience, software, and most of all, mobility. But they weren't compelling or ambitious. They weren't nearly enough to remake Ford or the public's perception of it, and never amounted to much anyway.

Meanwhile, the sort of bureaucracy that seemed so natural in Detroit crept into Ford's Silicon Valley outpost. Employees who'd had company credit cards were told to file expense requests in a spreadsheet that someone would review once a week. That didn't feel like a long time for a business that spent five years developing a vehicle, but it was agonizing for employees used to software cycles measured in months. It was the sort of corporate indolence that Silicon Valley detested. Worse, most of the work done in the lab amounted to little or no use, and was often ignored by the people who made the decisions in Dearborn. Apart from logging a few hundred miles on the I-10 freeway in March of 2016, Ford never tested its cars in California. The Palo Alto team played a lot of pickup basketball and worked from home on Fridays. Many who'd signed up to help an American titan get its groove back left for nimbler outfits that paid better. One engineer had started his career with a major automotive supplier, where his boss warned him that with its years-long timelines and endless testing, working in the auto industry was like "a long, slow donkey ride through hell." After less than two years with Ford, he dismounted and went to work for one of the small, autonomy-focused startups that were popping up all around Silicon Valley.

Ford's torpor, especially when it came to developing self-driving technology in-house, stood in stark contrast to what the Chauffeur team had accomplished in a few short years. But the Firefly car had taught the Googlers a lesson: Building a vehicle was a complex, intricate process in which they had negligible expertise. By 2015, Google had spun up a fair amount of hardware, including phones, tablets, and laptops. But a car that needed to stand up to the rigors of life on the road, where a single component failure could spell serious injury or death, required a different level of engineering. This was what the denizens of Detroit called "automotive grade," a catchall phrase meaning hardware that could take on cold, heat, rain, snow, potholes, and more, for tens of thousands of miles and years and years, and keep on rolling.

As they moved ever closer to a commercial launch of their robofleet, Chris Urmson and his teammates realized that they needed a major manufacturer on their side. Someone with the ability to crank out vehicles and help them integrate their sensors and computers into their vehicle instead of slapping them on as a retrofit, and who would grant them access to the software that ran the engine and other systems. And while Ford was lackluster in autonomy, poor at infotainment, and far from the bleeding edge of technological wonderment, it did build millions of automotive-grade cars and trucks a year.

In the spring of 2015, the two companies started talking about a partnership. They negotiated throughout that year, and the talks became a focus for John Krafcik when he came on as Chauffeur's CEO in the fall. The details shifted, but the basic outline was that Google would get its cars, and Ford would get a boost in the push toward autonomous driving. By December, the press was full of rumors about an imminent partnership, likely to be announced at the next month's CES. But the deal never came together. Postmortem accounts painted Mark Fields as overeager for a big, flashy deal. Something that would lend credence to his efforts at modernization, especially in the eyes of Ford investors reading reports about a changing industry and seeing the value that companies like Uber were producing. Google was less interested in that kind of collaboration. In early January of 2016, it pulled away from the deal.

When Fields took the stage at CES a few days later, he had no grand partnership to announce. Instead, he talked about tripling Ford's fleet of self-driving development cars, from ten to thirty Ford Fusions. In May, Google announced that it had struck a deal with Fiat Chrysler to install its tech in a fleet of Pacifica hybrid minivans. "The deal is the most prominent example yet of a Silicon Valley company collaborating with a traditional automaker on self-driving vehicles," the *New York Times* wrote, in exactly the sort of story Ford had hoped for. "It could also prove to be a breakthrough in the generally wary relationship between technology and auto companies, and prompt more collaborative efforts."

However strange her car-bound colleagues found it, General Motors Engineer Alisyn Malek biked to work every day. She lived about a mile from the office, and cycling was faster and easier than getting in the car just to drive a few minutes and park several blocks away. Even in the depths of the Detroit winter, she'd bundle up in waterproof gear, leaving just her eyeglasses poking out, pedal over, and lock up her bike at one of the few racks outside the Renaissance Center, the cluster of seven interconnected skyscrapers that was home to GM.

The Detroit native, in her mid-twenties, with a head of long red hair, wasn't the typical GM employee. She was an engineer who indulged her artistic side. In her free time, she made collages infused with progressive social commentary. She played saxophone in a band called Botanical Fortress. She and her woodworking husband ran an artist collective called Corktown Studios, where local creatives could work and display their creations. She lived in Detroit proper instead of the city's sprawling suburbs. And she didn't take for granted the superiority of the human-driven, gasoline-powered car.

"We've grandfathered in that it's completely sane for a human being who has gone through a rudimentary test to manage a combustion process in a 2,000-pound piece of metal," Malek said. "We're totally bought in on that." And after a century of that buy-in, which GM had done so much to encourage, she was in a wing of the company pushing people to learn a new way. She had started with the automaker as an intern in 2008, while working on a master's in energy systems engineering from the University of Michigan, landing an internship with the team developing the Chevrolet Volt. The plug-in hybrid was a new kind of car. Unlike the Toyota Prius hybrid, it could drive up to thirty-five miles using its battery alone, creating no CO_2 emissions. Unlike purely electric cars, it could supplement that limited range with a small gas-powered engine. The company created the Volt partly in response to increasingly strict government fuel economy standards, and partly to grab some of the good headlines Toyota had earned with the eco-friendly Prius. GM also wanted a riposte to a big-mouthed upstart named Elon Musk, the billionaire head of Tesla Motors, who was promising a luxury electric sports car.

The creation of the Volt was quite different from the sort of engineering Malek had expected at old GM. The electric power train raised all sorts of new questions. Along with technical issues like how lithium-ion batteries would degrade over time and how to ensure they didn't combust, the shift away from gasoline forced GM to rethink the customer experience. All of a sudden, its job didn't stop when the driver pulled away from the dealership. Getting a customer into an electric car meant going into their home (if not literally), helping find an electrician who could hook up a new high-voltage system, explaining when to plug in and for how long. It meant convincing people that abandoning the gas-powered system in which they'd been raised was really worth the trouble.

At the end of her internship, Malek became a part-time student and full-time GM employee. After Chevrolet delivered the Volt two years later, she spent a few years with the electrification folks and led the engineering team integrating ethernet connections into GM cars. In early 2014, she heard about a new arm of the company. Called GM Ventures, its job was to go around the world passing out money to people with interesting ideas. "How is that a job?" Malek wondered.

GM Ventures was the creation of Steve Girsky, one of the change-inclined outsiders who had taken over the company in the summer of 2009. GM had suffered from all the same problems as Ford—mismanagement, risk aversion, hubris, crummy products—but lacked the wherewithal to make major changes before the financial crisis struck in 2008, which decimated car sales and sealed the 101-year-old American giant's fate. When GM declared bankruptcy, President Barack Obama bailed out the giant. One condition of the deal was the installation of new leadership to loosen up the automaker's sclerosis. Girsky, a former Wall Street analyst, was the new head of strategy. He saw GM as the latest Kodak or IBM, companies that saw their dominance eroded by their inability to adapt to changing trends. Girsky figured a venture capital arm would force GM to take some risks, look at fresh ideas, and learn a few tricks from the West Coast of the country about how to bring new ideas to life. "The Silicon Valley way is *Hey, let's throw it out there. Shoot, correct. Reshoot, correct*," said Byron Shaw, the Ventures team's first lead. Detroit was more like "*Aim, aim, aim, aim, aim, aim. And finally, yeah, we're gonna get around to shooting at some point.*"

The in-house ventures unit was usually just four or five strong, typically staffed by what GM called "high pots," employees with the potential to rise to leadership. They were charged with finding ways to improve the myriad parts of GM's vast business. Malek attended conferences, spent time in Silicon Valley, and met many startups. When she was in Detroit, she talked to engineers, asking them where they needed help or new capabilities. Then she'd look for investments that might deliver. About a third of the GM employees she talked to were intrigued. Another third wanted nothing to do with the Ventures unit. The final third were the folks so into the idea that they were bringing ideas to her. That last group was the one that included Jim Nickolaou, who was working on a new feature called Super Cruise.

Nickolaou was the Carnegie Mellon alum who, as a GM engineer, had helped secure victory in the 2007 Urban Challenge by setting up Boss's transmission to accelerate as aggressively as possible, and he gave his first place medal pride of place on his mantel. His crew worked out of GM's Warren Technical Center, twenty minutes north of downtown Detroit. The 710-acre campus had hosted GM's most advanced engineering and design work since it opened in 1956. Built for $100 million (close to $1 billion today), it embodied an era in which GM was thriving. The era when it presented the Firebird II, the concept car that would "literally drive itself" by following wires buried in the pavement. Warren's glass buildings, wide lawns, and bucolic lakes and fountains were designed by Finnish architect Eero Saarinen, who would go on to create the famed TWA terminal at John F. Kennedy Airport in New York and the Gateway Arch in St. Louis, along with R&D centers for IBM and Bell Labs. *Life* called Warren "a Versailles of Industry." And even if twenty-first-century GM wasn't a power to match the Sun King, Nickolaou's team was doing some rather impressive work there.

After the Urban Challenge in 2007, Chris Urmson and Red Whittaker had approached GM R&D chief Larry Burns about pushing ahead with autonomous driving research. Given the automaker's dire financial situation, Burns had declined. The leaders who took GM through bankruptcy mostly protected the R&D budget, but weren't about to create a Chauffeur-like program. For one thing, they couldn't match the hundreds of millions of dollars Google was spending on its effort. GM had little to spare for a technology that most assumed was decades from reality, and whose near-term utility to the auto industry was hard to imagine. If a robo-car could only handle certain parts of the world—as Google and Uber anticipated—who would buy it? Moreover, people liked to drive. It was part of the American way, a deep-rooted part of the country GM had helped to build into a superpower. GM would continue its slow and steady approach to technological development, building ever more capable cars. If Chauffeur was launching a moonshot, the Detroit automaker was climbing a ladder.

Super Cruise was the next rung. The feature, destined for GM's luxury brand, Cadillac, would look very much like the highway driving system that the Chauffeur team had developed. On designated stretches of interstate, it would keep the car in its lane and a safe distance from other vehicles. Drivers could take their hands off the wheel and feet off the pedals, but had to keep their eyes on the road, and remain ready to retake control if the car ran into trouble. What Google had ultimately dismissed as a half measure looked tailored for GM. Unlike the tech giant, it had an existing auto business, built around a century of steady improvements, model year by model year. If it was ever going to remove the human from the act of driving, it would get there gradually. A car that could handle the highway might not help everyone, but for those who could afford it, driving would be more relaxing and likely safer. And it would keep GM competitive with Mercedes and others, who were developing similar features.

Like everything GM did, Super Cruise had to meet the standard of "automotive grade." It had to work for years and tens of thousands of miles. And even on a luxury car, it had to be cheap—no more than a few thousand dollars. Those demands meant relying on radar, which was cheaper and more durable than Lidar laser scanning, and didn't involve a designer having a conniption when she heard some engineer wanted to stick a spinning coffee can on a car's roof. The problem with radar—the problem that made Lidar key to Google and Uber's efforts—was that it offered no more clarity than a Rorschach test. A car that couldn't tell the difference between a highway sign and a Mack truck was useless. So the GM team programmed their software to lean on radar's strong suit, determining the speed of the objects it detected. It would pay attention only to moving things, which were likely to be vehicles. This simple and logical hack, though, meant that if the car came across a stopped vehicle, like a police cruiser or construction truck, it might not "see" it. And it might not hit the brakes.

Nickolaou's team wanted to take the driver's hands off the wheel, but keep his eyes on the road. Google's experiment with the highway made clear that once the computer was doing the driving, the human was unlikely to stay alert, even if he tried. To immunize Super Cruise users, the GM engineers stuck an infrared camera, the size and shape of a gumdrop, behind the steering wheel. It would watch the driver's head and eyes, noting when she gazed out the window, turned to her passenger, or looked at her phone. If she did so for more than a few seconds, the car would issue an audio alert. And if the driver didn't perk up after a few warnings—perhaps being stubborn or incapacitated—the car would put on its hazard lights and slow to a stop.

Nickolaou worked on another way to bolster the system, not so different from how Urmson made the Red Team's original Sandstorm a better racer: Super Cruise would use maps. High-definition maps, made with Lidar scanners and accurate within a few inches, of every bit of divided highway in the United States and Canada. One hundred thirty thousand miles in all. The benefit was twofold. Advance knowledge of lane line location would help the car stay centered even on sharp curves. And by restricting where it would engage, risk-averse GM limited the opportunity for someone to misuse the feature. Beyond keeping its use to divided highways—the type with on and off ramps, and no intersections—the team could cut out toll plazas, exit only lanes, and anywhere else that involved making a decision. Like the Chauffeur team, GM would make its autonomous feature work by skipping the trickiest bits. But GM didn't have those maps, and Nickolaou didn't have access to Google's fleet of Lidar-topped cartography cars. So he went to Alisyn Malek for help. Malek started asking around Silicon Valley, looking for anyone who might have a way to create a high-resolution record of every foot of interstate in the US and Canada, and then Europe and China. She eventually found the answer in an outfit called GeoDigital, which traversed the country with laser-equipped cars and planes. But amid her search, Malek came across a company that would prove to have a far greater impact on General Motors' future.

At the end of 2013, General Motors had announced another change in leadership. Four years after filing for bankruptcy, the automaker was leaner, smarter, and more efficient thanks to the guidance of outsiders like Steve Girsky. It was time to hand the company back over to the natives. The new CEO would be Mary Barra, the first woman to lead a major automaker. The daughter of a man who spent forty years in a Pontiac factory, Barra had joined GM at eighteen, graduating from the General Motors Institute (now Kettering University). After two years in the CEO job, Barra made, if not the boldest declaration about the molting nature of her industry, the most significant, by virtue of her position. "I believe we'll see more change in the automotive industry in the next five to 10 years than in the past 50," she wrote in an essay titled "The Year Detroit Takes on Silicon Valley," published on LinkedIn in December 2015. "Driving this historic transformation are shifting views of vehicle ownership, growing urbanization and the very digital and sharing economies that have disrupted so many other industries." She talked about shared mobility, autonomous tech, and electrification. And she made clear that GM was getting ready. "I have committed that *we* will lead the transformation of our industry," Barra wrote, jabbing at Google and Uber. "No matter how our customers chose [sic] to get from one destination to another, we will provide them with choices that make their life better and easier."

Leading into January 2016, while Ford watched its deal with Google disintegrate, GM made a flurry of moves that backed up Barra's bold talk. It acquired what was left of Sidecar. The early ridehail business had been pulverized by Uber and Lyft, but brought valuable intellectual property and experienced employees into Detroit. GM launched a car-sharing service called Maven. And it announced a new partnership with Lyft.

The automaker would invest a massive $500 million in Uber's archrival, take a seat on its board of directors, and provide cars to some Lyft drivers via short-term loans. The news that the headlines focused on was that the two new partners planned to jointly create a self-driving ridehail service. Another competitor would now join the fully autonomous space that Google had pioneered and Uber had leapt into, this one coming from Detroit. To make it work, Barra's company just had to figure out how to build a self-driving car.

Super Cruise was scheduled to debut in 2017 as a $2,500 option on the Cadillac CT6 sedan, and spread to other models from there. It would mark a technological leap forward for GM, the industry's first commercial product that encouraged drivers to take their hands off the steering wheel. But it wasn't the right foundation for building the technology that a taxi-like system would require. As Chris Urmson had told the audience at his March 2015 TED talk: "Conventional wisdom would say that we'll just take these driver assistance systems and we'll kind of push them and incrementally improve them, and over time, they'll turn into self-driving cars. Well, I'm here to tell you that's like me saying that if I work really hard at jumping, one day I'll be able to fly."

GM needed a fresh approach. And since its tech development cycle worked at the pace of a stone engraver—Super Cruise development had started in 2008—it needed someone with the skills of a stenographer. GM didn't have decades to catch up with Google and Uber. It had investors who wanted to see it get on its horse, because they were reading the analyst reports predicting a major disruption to the auto business. And if they wanted to see what an automaker with a Silicon Valley ethos could be worth, they just had to glance at the front page of any tech news site for the latest Tesla story.

In a time when Americans were far from impressed with home-grown automakers, Elon Musk's car company had legions of devoted fans, even among those who couldn't afford its products. Musk also ran SpaceX, a company making reusable rockets, and was the model for Robert Downey Jr's portrayal of Tony Stark, the alter ego of superhero Iron Man. He engaged with fans on Twitter, taking their suggestions, floating ideas for colonizing Mars, and making up new concepts for su-personic tube-based travel. Tesla's electric cars went hundreds of miles between battery charging stops. Spine-bruising acceleration made them a match for Ferraris and Lamborghinis on the drag strip. Wireless soft-ware updates made them smarter and more efficient, even years after they had left the store. And in October 2015, one of those upgrades gave them the ability to drive themselves.

About a year before, Musk's Palo Alto–based automaker had started equipping its cars with the cameras and radars they would need to find lane lines and stay a safe distance from other vehicles. The update also delivered the software that put the equipment to use. Tesla had started work on Autopilot long after GM had begun developing Super Cruise, and beat it to market by nearly two years by creating a simpler system. It eschewed the maps that GM's cars used to hem in Super Cruise's abilities. Instead of an infrared camera, it settled for the simplest way to gauge driver awareness, a sensor that detected when the driver touched the wheel. Tesla told its customers they should keep their hands on the wheel, but when Musk and his employees demonstrated Autopilot, they put their hands in their laps. (GM, by contrast, included warnings about paying attention even in press releases about its system.) Within a week of the release, drivers had uploaded videos to YouTube of the cars swerving into traffic and taking an exit when they should have stayed on the freeway. A trio of drivers used the self-driving feature to set a cross-country speed record, setting Autopilot to 90 mph or more and driving from Los Angeles to New York City in fifty-seven hours and forty-eight minutes. Musk, who had told his customers "to exercise caution in the beginning," sent them a congratulatory tweet.

But then, in May 2016, Tesla hit a grim milestone. Former Navy SEAL and Tesla superfan Joshua Brown was driving his Model S in Autopilot down a rural Florida highway when a semitruck turned left across his path. Without slowing down, the Tesla hit the semi and went right under it, cleaving off its roof. It kept going, off the road, and three hundred feet later hit a utility pole hard enough to break it. The car went another fifty feet before stopping. Brown died, the first publicly recognized victim of a self-driving car. The US National Transportation Safety Board found that while Brown was inattentive "due to overreliance on vehicle automation," Tesla bore some of the blame for designing a system that allowed that inattention. The report, issued in September of 2017, angered Musk but didn't slow the financial world's lust for Tesla. That same month, the Silicon Valley automaker's stock price approached an all-time high. According to Wall Street, Tesla, which made only a few thousand cars a year, was worth more than the Ford Motor Company. At one point, its valuation even surpassed GM's.

Tesla was an example of what a high-tech reputation could mean. Efforts like GM Ventures had drawn the centenarian automaker into Silicon Valley, but in early 2016 GM was realizing that a self-driving service to rival Google and Uber would make the real splash. And that's when the company's leaders heard more about a startup called Cruise Automation.

Kyle Vogt was thirteen years old when he started thinking about autonomous vehicles. He'd gone to Las Vegas to participate in *BattleBots*, the show about fighting robots. After his machine, a dustpan-shaped thing called Decimator, was itself decimated by the pickax-wielding Tazbot, Vogt and his father drove back to their home in Kansas City, Kansas. On the straight slog across the American plains, Vogt had a thought about driving on the highway: "This seems like something a robot could do."

As a freshman at MIT, Vogt joined a team competing in the 2005 DARPA Grand Challenge, but the undergraduate effort was eliminated in the qualifying round. He dropped out during his junior year to co-found a non-robotic venture called Justin.tv. That evolved into Twitch, a hugely popular streaming service that let users watch strangers play video games. Vogt was a driven, talented coder. "He'd just, like, lock himself in a room for three days and code away and then emerge with something that worked," a teammate said. And he had a nose for business. He and his cofounders sold Twitch to Amazon for $970 million.

By then, Vogt had returned to the idea of self-driving cars. In September of 2013, when Google's Chauffeur team still looked to be the only outfit pursuing the technology, he founded Cruise Automation. The San Francisco–based startup began by developing an aftermarket kit that used cameras and radars to give a conventional sedan the ability to drive itself on the highway. Cruise showed off a prototype in June of 2014, and aimed to have a product on sale the next year for $10,000. But the next fall, Vogt decided to chase a greater goal: Instead of settling for a highway system, he would go head-to-head with Google in the race to develop a self-driving car that could take the human entirely out of the equation.

It was during her search for a company that could help make maps for Super Cruise that GM Ventures' Alisyn Malek met Vogt through a venture capitalist they both knew. Vogt couldn't help with the maps, but they stayed in touch. And when General Motors announced in January 2016 that it would work with Lyft to create a network of self-driving cars, Malek passed Vogt's name up the ladder.

That deal with Lyft was the work of Dan Ammann, who'd come to the automaker from Morgan Stanley to help guide it through the process of going public again after bankruptcy. A burly fellow with a deep voice who'd grown up on a New Zealand dairy farm, Ammann became GM's president when Mary Barra took the CEO job in early 2014. The new leadership team spent its first year dealing with the fact that GM had built bad ignition switches into cars for years. The faulty devices could shut down the engine while driving and deactivate the airbags, and were linked to 124 deaths. GM ended up recalling nearly 30 million cars around the world and paying $900 million into the US Treasury. It was a rude opening salvo, but just the beginning.

"We put our heads up, and said, 'OK, everything's under control,'" Ammann said. But while he, Barra, and the rest of the company had been finding their footing, Americans had been finding new ways to get around. Uber and Lyft still represented a tiny fraction of the transportation market, and weren't going to much affect the bottom line for GM, which made most of its money by selling pickup trucks and SUVs to suburban and rural customers. But Ammann registered that consumers were voting with their dollars on the new services. He started making trips out to Silicon Valley, meeting with various players in the changing transportation space. It was in a sit-down with Lyft that the opportunity to expand GM's business struck Ammann. The ridehail company's leaders told Ammann that if GM or anyone else could build a fully self-driving car, they'd pay big bucks for it. "Well, shit," Ammann said to himself. "That's pretty interesting." Yet by the time he formed the partnership with Lyft, pledging to create a robo-taxi, he knew that GM needed help to make that happen.

The automaker's in-house R&D team had done impressive work. Super Cruise was an excellent system, but it was the result of nearly a decade of development. A totally autonomous car was a whole other beast. It required different sensors—including a Lidar laser scanner—and software that was orders of magnitude more complex than the sort that guided a car down the highway. Ammann knew that even if they could match what Google was doing, they couldn't do it anytime soon.

By 2016, Cruise was far from the only self-driving car startup in Silicon Valley. Lots of talented engineers had seen the hype surrounding the technology, especially after Uber jumped in. But when Ammann added Vogt to his list of people to visit in California, he was impressed.

The young engineer had built a team of forty people and in just a few months had developed a prototype car that could handle some of San Francisco's wild streets. In January and February, Ammann made several visits. Each time, the car was more capable, executing more complex maneuvers more smoothly. And Ammann really liked that Vogt kept their meetings short—explaining he had an enormous amount of work to do and was eager to get back to it.

Ammann, too, was eager to get moving. In late February, he flew out to San Francisco once more. He met with Vogt and talked about how best to work together. On Friday, he flew home to Detroit. By Monday, he had started drafting the terms under which GM would acquire Cruise Automation. Ammann had to sell the idea to GM's board—he intended to spend a lot of money on a company that had just forty employees and no actual product—but by the end of one meeting, he had them convinced. GM in its heyday had been all about dominance. But, Ammann's math said, if it could account for just 3 percent of vehicle miles around the world, it would be worth more than Amazon or Apple. It would be a true giant once more—and it would make its investors extremely happy.

On March 11, 2016, General Motors announced that it had acquired Cruise for nearly $600 million. Vogt and his team would stay in San Francisco as a relatively autonomous unit, charged with turning their rough prototype into something that could carry America's biggest automaker into the future.

On a sunny Tuesday morning in August 2016, a gaggle of reporters surrounded Mark Fields. They fired questions and scribbled notes as the Ford CEO explained what he meant when he said that come 2021, Ford would deploy a fleet of fully autonomous vehicles as part of a ridehailing service. Fully autonomous, as in no steering wheels, no pedals. They'd start off in one to-be-named American city, then expand to cover more territory. And by "fleet," he meant hundreds of vehicles.

Fields had come to Ford's research center in Palo Alto to make this announcement, the first time the automaker had made clear an intention to shoulder its way into a race dominated by Google, Uber, and now GM's Cruise. "We see the upcoming decade for the automobile really centered around the automation of the automobile," Fields said.

After the scuttled deal with Google, he had made a series of moves to prepare Ford for that decade, investing in a high-resolution mapping company and Dave Hall's Velodyne Lidar company, and forming a deal with a machine learning outfit. A few weeks after the announcement, Ford acquired Chariot, a San Francisco startup using vans to transport commuters who didn't live near (or like) public transit. This was less a revenue-generating move than a longterm play to investigate what it would take to operate a fleet of shared vehicles.

John Casesa knew it wasn't enough. The former analyst had been Ford's head of strategy for more than a year by this point, long enough to see that the automaker's normal research-and-development team wasn't up to the technical aspect of what Fields had promised, certainly not with a five-year deadline. Yes, the team had been working on autonomous tech since the 2005 DARPA Grand Challenge, but Ford's leadership had never taken that work seriously. While Google had spent years pumping money and resources into a fully autonomous vision, Ford had directed its funding and manpower toward other priorities, including staying alive through the American auto industry's darkest days. It didn't even have a semi-automated driving system with the kind of capability that Tesla's Autopilot or GM's Super Cruise could offer on the highway. The challenge of creating a truly autonomous car that could handle the complexity of an urban environment, Casesa realized, was beyond the skills of an organization that had spent the past century stamping sheet metal. "We weren't a software company," Casesa said. "So we were going to create a software company."

A company that would exist outside the normal Ford hierarchy, where it would be safe from the corporate giant's tendency for bureaucracy and inertia, where it could function something like Google's

Chauffeur team. A company that would exist far from Detroit itself, where it could pull in the computer science talent it would need to have a chance to make good on the 2021 pledge.

Ford's timing, at least, was good: Around the time Fields made his promise, Bryan Salesky was looking for a new gig. Since helping Chris Urmson win the 2007 Urban Challenge for Carnegie Mellon and General Motors, Salesky had developed autonomous mining machines for Caterpillar. He had joined Google's Chauffeur team, then quit a year later, frustrated by the tension between Urmson and Anthony Levandowski. Urmson got him to rejoin in early 2013, and to take Levandowski's job as hardware lead. Salesky struck a balance between Chauffeur's two leaders: He moved deliberately, but stayed focused on the goal of putting a product into the marketplace. By September 2016, he was among the many Chauffeur engineers looking for something new, where he would have final say on how things got done. He had talked to Uber about joining its effort before the ridehail company decided to move ahead with Levandowski.

Salesky had always been more attached to the eastern half of the country, having spent his childhood in Michigan before moving to Pittsburgh for college. After working on the Firefly car, he thought the right path forward was to work closely with an automaker, something Google resisted. He'd seen the challenges of building a vehicle from scratch, but also the upsides of being able to design one that accommodated a robo-car's sensors and robust power requirements. He teamed up with Pete Rander, who had been his boss when he started at Carnegie Mellon's NREC lab a decade earlier. Rander had been part of the group that jumped ship to Uber's self-driving effort, and quit the ridehail company around the time Salesky left Google. In the fall of 2016, they quietly incorporated a company called Argo AI. They thought that an outfit with serious software skills that paired with a big automaker would provide the right combination of strengths necessary to put an autonomous vehicle into the marketplace. "I didn't think that entity existed," Salesky said. "So we put that hook in the water." Casesa bit, and offered to make Ford Argo's major, and sole, investor. Salesky and Rander agreed.

This arrangement wasn't an easy sell in Detroit. A major investment in an outside company was a clear signal that Ford was behind in the self-driving race, that its leadership had been faffing about while the rest of the industry steamed forward. But Fields was for it, especially after seeing his hopes of collaborating with Google dashed the year before.

And so, in February of 2017, Ford announced that it was investing $1 billion over five years in a startup that nobody had heard of until that moment. Argo AI would be based in Pittsburgh, have its own board of directors, be able to offer equity to employees (crucial for attracting top software talent), and keep the right to license its software to whomever it liked. Ford, its main stakeholder, would be responsible for building the vehicle itself, taking an expensive and pesky hardware problem off Salesky's plate.

If GM's acquisition of Cruise and its forty employees looked like a prayer, Ford's move, a year later, into the rapid evolution of a new industry looked more like a Hail Mary pass. But the general impression among people who knew Salesky was that if anyone could build the team that would give Ford the twenty-first-century equivalent of the Model T—the self-driving car for the masses—Salesky was a good choice. Even Levandowski, who had lost his gig as Chauffeur's hardware lead to the new Argo CEO, concurred: "I wouldn't bet against him."

A Death in the Desert

A LITTLE AFTER NINE IN THE MORNING ON A SUNNY SATURDAY in October 2017, Chris Urmson stepped onto the green rectangle known as the Mall at the center of Carnegie Mellon University's campus in Pittsburgh. Wearing blue jeans, canvas sneakers, and a subtly checkered sport coat, he parked the black carry-on suitcase he was rolling behind him on the paved path and walked over the carefully manicured grass toward the four vehicles standing in a line.

The quartet stood in front of Hamerschlag Hall, which was called Machinery Hall when it was built in 1906, one of the first buildings to house a new school dedicated to the pursuit of technology. Pittsburgh steel magnate Andrew Carnegie wanted a campus whose layout resembled that of an "explorer's ship." This arching, cream-colored stone building, an emblem of the era's City Beautiful and Beaux Arts movements, was the helm. Its defining feature was the rotunda that resembled a crown but was in fact a well-disguised smokestack. At the dawn of the twentieth century, the university was too far from downtown Pittsburgh to connect to the city's power grid, so it had to produce its own electricity. A century on, the rooms where every freshman had been required to spend two weeks shoveling coal to power the school had been converted into "clean rooms," free of any particles or dust that could interfere with the computer chips and hardware that students now created there. This evolving home of technological wonders was the perfect backdrop for the day's photo op.

The oldest of the four vehicles was Terregator, the six-wheeled mine exploring and mapping robot that looked like a desk and moved about as fast as one. On its right was H1ghlander, the cardinal-red Humvee that had come up limp in the desert outside Primm, Nevada, in the 2005 DARPA Grand Challenge. To its left was Boss, the Chevrolet Tahoe with the huge GM logo on its hood that had delivered Carnegie Mellon's long delayed victory in the 2007 Urban Challenge. Urmson walked past them all, drawn to the vehicle at the end of the line, the original Red Team racer. The one for which he had left the Chilean desert, the one that had sent him into the Mojave and then onto a lifelong quest to render possible the impossible.

He ran his hand along the machine's rough metal edges, feeling the damage from the crash (*Too fast! Too fast!*) that almost sank its chances in the first, 2004, Grand Challenge. Circling around to its left side, Urmson moved past the steel square that passed for a seat, where he had spent so many hours with his hands hovering over the go-kart wheel, his right side aching from the weight of the massive electronics box pressing into him every time the vehicle turned. Standing in front of the box, Urmson undid a pair of beaten-up latches and flipped open a small hatch. He twisted his body to reach his hand—one of its nails permanently chipped by the Humvee's cooling fan fourteen years earlier—past where his eyes could see. As his fingers confirmed that the circuit boards he had programmed to conquer the desert were still there, Urmson grinned and let out a small "ha!" Sandstorm still had its heart.

Under his sport coat, the Canadian engineer wore a purple T-shirt reading "Aurora," the name of the self-driving car startup he'd launched in 2016, a few months after stepping down from the Google team. His cofounders had led early autonomy efforts at Uber and Tesla, respectively. The trio framed Aurora as a fresh start built on what each of them had learned, and took an open-minded approach to the business. They kept Aurora independent but built a collection of automaker partners, including Hyundai, Kia, and Byton, a Tesla-like, all-electric startup that knew autonomous driving was a must-have feature, especially for an industry newcomer. Aurora's plan was to work as a sort of supplier, building what it called its "Driver" for anyone who was interested. Urmson spent most of his time in Silicon Valley, and had headed east to check in on the company's Pittsburgh office, and for a reunion. It had been a decade since Tartan Racing's Boss won the Urban Challenge, the culmination of a years-long effort that pushed Urmson and many others to work harder than they ever had before, harder than they'd thought possible. Carnegie Mellon was marking the anniversary with a small conference, a chance to get the old team back together and to talk about the emerging world of self-driving cars that all their sweat and stress and pain had created.

Urmson spent fifteen minutes or so on the lawn, inspecting the robots. He told stories from his days with the Red Team and explained various features to the few people around, including three of his own employees, young enough to have been in grade school when their boss was cracking the code for a challenge he thought was unwinnable. He had long since been vindicated. His job now was to deliver that code to the public, and to wrestle down the pesky details that came with any new technology: operational efficiency, regulations, insurance, business plans. He was far from the only one seeking the answers.

The day's festivities about to begin, Urmson led the group to the engineering building reserved for the get-together. He held open doors for everyone as he navigated by muscle memory the campus where he had spent so many years. Where he would likely have been content to spend many more, had Sebastian Thrun not drawn him to the West Coast.

The lecture hall's seats, their upholstery stitched with tiny zeroes and ones, were filled with familiar faces. Tony Tether had driven up from Washington, where he worked as a consultant. One of his clients was a Lidar company with scores of competitors in what had become one of the hottest fields in tech. Less than a year into his deal with Ford,

Argo CEO Bryan Salesky was getting Argo off the ground and onto the streets of Pittsburgh. Kevin Peterson of the Red Team was visiting from San Francisco, where he ran a company building robots that rolled along sidewalks delivering food. GM's Jim Nickolaou had made the three-hundred-mile drive from Detroit in a Cadillac CT6, enjoying the Super Cruise system he'd helped design.

At the center of the celebration, of course, was Red Whittaker. Approaching his seventieth birthday, the former Marine still scoffed at the idea of relaxing on weekends, let alone retiring. He'd spent years working on what became known as the Google Lunar X Prize, an open challenge to land a robot on the Moon and have it travel five hundred meters, beaming data and images back to Earth. Whittaker hadn't made it happen. Neither had anyone else, and the competition would be officially ended in early 2018—a sign that not just any open competition could spark innovation. DARPA, on the other hand, had built off the self-driving challenges with a series of similar prize-based contests addressing robotics, cybersecurity, predicting infectious disease, and more.

The lunar loss hadn't slowed Whittaker. He was already going hard on a new generation of nuclear reactor robots. He had found the time, though, to play emcee for this event, which included a celebratory dinner and a series of panel discussions covering robotics history and the future of autonomous vehicles. During dinner, Whittaker took the opportunity to resolve a mystery that had troubled him and many others since October 8, 2005: What had happened to H1ghlander that day in the desert in the second Grand Challenge, when it gave way to Stanford?

While booting up the old Humvee to move it to its place on the lawn, Whittaker had brushed against a small black box, and heard the engine slow. That box modulated how much power ran to the engine. He realized that when H1ghlander had flipped during that practice run just before the 2005 Grand Challenge, that module had been damaged, so anytime something knocked against it, it would cut the engine's power to nearly nothing. That's why H1ghlander hadn't just stopped on the roads outside Primm, but slowed and accelerated at an inexplicable rhythm, until Sebastian Thrun's robot left it behind. The Red Team's lengthy autopsy had focused on the sensors, the code, the work they had done. They missed this little thing, one of the electric components that came stock with the military truck. Announcing the discovery, Whittaker held the black box up in his hand. "How about that, buddy?" he said to Urmson, slapping him on the back. "You're off the hook!"

In 2005, that malfunction had been a source of tremendous pain. By 2017, it had morphed into an almost tender memory, the way you might think of the high school sweetheart who dumped you. Much had changed in those years, and many of the onetime teammates filling those lecture hall seats were now competitors, motivated as much by the interests of investors as by any thrill of problem solving.

The seeds planted by the DARPA Challenges had burst into an ecosystem. Gone were the doubts that autonomous driving would reshape daily life; they'd been replaced by the central question: Who would gain the most from delivering this technological shift, and who would suffer its fallout? Every major automaker and most serious tech companies were pursuing autonomous driving in some way. Countless startups were seeking funding to create their own software or fill some niche. National legislation and regulations were in the pipeline. Unions were bracing themselves for yet another technological wave that could wash away the jobs that had helped build large swaths of America's middle class. The basics of the software were so established that building a self-driving car was a common project for college-level engineering students. Sebastian Thrun was using Udacity, his online education company, to train thousands of self-driving engineers a year. In 2016, when he first offered the course, thirty-five thousand people said they were interested in enrolling. Dave Hall had opened a massive manufacturing plant in San Jose, building ten thousand Lidar sensors a year and hoping to fend off the scores of competitors joining the industry he'd created. The American military had come nowhere near meeting the mandate that had sparked the Grand Challenge, to make one-third of its ground vehicles unmanned by 2015. But it, too, continued to pursue the technology.

It was impossible to know what that chase would produce, but the development of the human-driven vehicle offered a lesson. When automobiles first hit the road, people called them horseless carriages, and for good reason: They worked just like horse-drawn carriages, without the clopping and the manure. But as auto technology evolved, they become capable of much more, and people started calling them cars. The common understanding of driverless cars—as taxis without the taxi driver, or personal cars whose owners don't have to hold the wheel—displayed the same lack of imagination. This latest technological shift was sure to produce changes nobody could foresee, and likely outgrow a name pulled from the paradigm it replaced. In just a few years, the experts said, autonomous driving would generate billions of dollars. It would redefine or vaporize the way one in nine Americans made a living, and unlock untold wealth for whoever could make that happen. At the end of 2016, Google's Chauffeur project had undergone a metamorphosis. It became its own company, a subsidiary of Alphabet, which Google had created as its parent company in a 2015 reorganization. And it was now called Waymo. As a final challenge before this step, and a proof of readiness, the team (minus most of the engineers who'd been there for the Larry 1K) had sent a blind man named Steve Mahan for a carefully planned ride around Austin, Texas, without a safety driver to intervene if the car made a mistake. The team had grown to hundreds of employees and logged 2.5 million miles on public roads, plus billions more in computer simulations. That growth had engendered multiple office relocations, and during one move, the original team's collection of ten empty, signed champagne bottles, hallmarks of their completing the Larry 1K challenge, had disappeared.

The explorers had undergone their own changes. They were settlers now, trying to establish footholds in what had become a proper industry. And not everyone was getting along with his neighbors. "It's a little sad to see what's happening in the industry today with some of the questionable ethical behavior that's out there," Urmson said. "I guess that's part of the evolution, part of how you tell this is immensely important and valuable. But for me, there's a little bit of nostalgia for when everyone was pulling in the same direction."

CMU's conference was heavy on nostalgia, but in conversations here and there, one could hear talk of the trouble Urmson was alluding to, always associated with a single name, the person who'd done so much to light the powder keg that DARPA's Challenges had filled.

In 2017, no one was toasting Anthony Levandowski. No one called him an "all-around whiz kid" or superstar anymore. Not after February 23, when Waymo filed a bombshell lawsuit against Uber, accusing it of a "calculated theft" of its trade secrets, violation of its patents, and all-around cheating to get ahead in the race to produce a self-driving car.

From the start, Waymo's argument boiled down to a simple accusatory narrative: A desperate Uber paid Anthony Levandowski an enormous amount of money to steal Waymo's intellectual property—chiefly its Lidar design—and use it to build a self-driving car for Uber. The suit asked for nearly $2 billion in damages, a sum that signaled how much value self-driving technology was poised to create. The consequences looked to be devastating. If Uber lost the case, it might have to pay Waymo a mountain range of money in restitution and kill the self-driving program CEO Travis Kalanick considered vital to long-term viability. If the Justice Department got involved and brought criminal charges, a scarier threat loomed. Trade secret theft could land Levandowski in prison.

So began a massive legal fight, as Uber's phalanx of lawyers lined up against Waymo's. In a San Francisco federal courthouse, Judge William Alsup dashed Uber's hope to move the case into private arbitration and rejected Waymo's bid to stop Uber working on its self-driving tech until the case concluded. But nobody seemed to doubt that Levandowski had walked out of Google with those downloaded files. "You have one of the strongest records I've seen for a long time of anybody doing something that bad," Alsup told a Waymo lawyer. He even took the extraordinary step of referring the case to the US Attorney's Office "for investigation of possible theft of trade secrets based on the evidentiary record supplied thus far."

Waymo, however, wasn't suing Levandowski. This was *Waymo v. Uber*, and for any charges to stick, Waymo had to prove not just that the files had left its servers, but that Uber had used them to its advantage. So followed months of discovery and depositions, as 129 lawyers sifted through millions of emails and technical documents for evidence of good or bad behavior, filing more than one hundred thousand pages of briefs, depositions, and motions.

Missing from all the back-and-forth over Levandowski's behavior was the man himself. Recognizing that he was at risk no matter what happened to Uber, Levandowski decided to use his Fifth Amendment right to avoid testifying and turning over any documents. As detailed in an Uber due diligence report later introduced into evidence, Levandowski had told Kalanick he had Google files before Uber acquired Otto, and said that he had destroyed the discs containing them. The investigators also found plentiful evidence of questionable behavior: Levandowski had fifty thousand Google emails on his personal laptop (he had synced his personal and work email accounts in 2014). His various devices contained Google videos, patent applications, and testing files. He had had an old Street View prototype camera and assorted robot parts in his garage (which he also had destroyed, he said). He made a habit of deleting his text messages and telling others to do the same.

Uber's lawyers argued that those files were of minor importance, but when it came to Levandowski's fate, that mattered little. The suit was threatening Uber's future, and Levandowski's refusal to testify made things worse. Kalanick still thought Levandowski was one of the world's best minds when it came to autonomous driving. But the CEO did what he always did: He moved to protect Uber. At the end of May 2017, he fired his self-driving superstar. Levandowski would get close to nothing from that reported $680 million Otto acquisition, because almost all of that money hinged on his team's hitting various milestones over the course of years.

Kalanick himself resigned as Uber's CEO just a month later, amid a swirl of scandals that included workplace sexual harassment and discrimination, psychological manipulation of the company's drivers, and deceptive practices in dealing with regulatory authorities. In the end, some of his biggest investors turned against him, even suing him for fraud, alleging that when he presented the Otto acquisition to his board of directors, he'd hidden the fact that Levandowski had taken data from Waymo.

When the lawyers finally went before a jury on Monday, February 5, 2018, the narratives they laid out in their opening arguments were simple. Waymo argued that Uber was desperate to catch up in the race to develop a self-driving car, and had schemed with Levandowski to close the gap Waymo had created by starting its research long before any other company realized the tech's potential. Uber painted a picture of Waymo as a market leader threatened by younger, faster competition, and willing to take extreme measures to ensure its engineers stayed within the castle walls. Whatever Levandowski had done, that was on him.

Over four days, as Waymo made its case, the jury heard testimony from Travis Kalanick, John Krafcik, and Dmitri Dolgov (who had taken Chris Urmson's place as Waymo's lead engineer), among others. But behind the scenes, the lawyers and the executives were negotiating. Waymo had smeared Uber's reputation, but failed to produce any slam-dunk evidence that the ridehail giant had profited from Levandowski's machinations. Dara Khosrowshahi, who replaced Kalanick as Uber CEO in August 2017, would be happy to defang one of the many potentially deadly scandals he'd inherited. So when the clock hit 7:30 on Friday morning—Judge Alsup liked to start early—the reporters squeezed into the courtroom's wooden benches were confused to find the lawyers and the judge missing. Twenty minutes later, Alsup took his seat, and gave the floor to a lawyer for Waymo, who moved to have the case dismissed, announcing that the parties had reached a settlement. Waymo would drop the case in exchange for 0.34 percent of Uber's equity—worth about $245 million—and a promise that Uber would not use any of Waymo's hardware or software in its self-driving cars. "All right," Alsup said. "This case is ancient history."

Levandowski had been silent since Waymo filed its explosive accusations. The press that had long been friendly to him started digging into his history, no longer satisfied with portraying him as the creative kid so eager to make robots real. *WIRED* revealed the details of Google's acquisition of 510 Systems and Anthony's Robots in 2011, the $20 million deal structured to reward Levandowski and not his employees. The same story noted that in September 2015, shortly before he was due to receive his $120 million bonus, Levandowski created Way of the Future, a church dedicated to the idea that as artificial intelligence progressed, machines were bound to rule over humans. "We're basically creating God," Levandowski said. And once humans shared the planet with something orders of magnitude smarter than them, there was no use trying to contain or control it. "It is for sure gonna get out of the cage." Way of the Future was a bid, he said, to show the computers that humans were on their side, to become pets rather than livestock.

Eight months after the trial's conclusion, the *New Yorker* ran a story replete with details that further tarnished Levandowski's reputation, including his "I Drink Your Milkshake" T-shirt and the argument with Isaac Taylor that led to the risky highway encounter with the Toyota Camry. The public perception of Levandowski went from talented, driven engineer to self-serving, irresponsible jerk. Even his nanny got in a kick, filing a bizarre lawsuit in which she claimed Levandowski had often failed to pay her, and recounted things she'd heard him saying, including considering fleeing to Canada when Waymo made its accusations. Levandowski called that suit "a work of fiction," but paid the nanny an undisclosed amount of money to drop it, which she did. One story from car news website Jalopnik, recounting the juiciest details from the *New Yorker*'s story, carried the headline "The Engineer in the Google vs. Uber 'Stolen Tech' Case Really Was Terrible."

Among many of his former colleagues, though, Levandowski maintained a peculiar kind of goodwill. "I like Anthony. I'm just afraid to have him around," Velodyne's Dave Hall said. Sebastian Thrun still marveled at his talent for getting things done and inspiring others. "He's very underestimated in the entire scandal. I think he's a way better person as a human being, and he's a way better executor than the way he's portrayed in the media," Thrun said, adding a careful caveat: "His attitude toward disclosure and truth is not anywhere near my ethical standards." And while plenty who worked with Levandowski were happy to see him go down—dismissing him as "a weasel" and "evil"—others echoed that two-part evaluation: *Anthony's got a lot of great qualities, but it was always going to end this way.* He had skated around trouble for years. Now, the ice buckled under his feet.

In a way, *Waymo v. Uber* was premised on an early incarnation of the self-driving world, when Google was the only game in town, and starting a viable competitor hinged on winning away its brainpower. By early 2017, the expert population had exploded, and the various contenders found their own ways to pursue an autonomous future. While Waymo and Uber clawed at each other, General Motors' Cruise went on to raise more than $7 billion by early 2019. Ford had Bryan Salesky's Argo AI. Chris Urmson started Aurora with a clean slate. Chauffeur alums Dave Ferguson and Jiajun Zhu launched Nuro, a startup focused on self-driving delivery robots. Don Burnette, who had cofounded Otto with Levandowski, left Uber to run his own robo-trucking effort, Kodiak Robotics. Zoox was raising billions of dollars with a promise to redesign not just the driver, but the idea of the car itself, developing a symmetrical, bi-directional custom vehicle. Alisyn Malek had left GM to help run May Mobility, making shuttles that drove themselves short distances along simple routes. Among them, they employed thousands of people, enough to make the fact that Google had started this industry with a handful of engineers in 2009 seem hard to believe.

The quest now was to prove not that cars could drive themselves, but that one could make a business out of it. That regulators and insurance agents and lawyers and most of all the public could be convinced that this technology was safe—and worth paying for. But just five weeks after Judge Alsup told the Waymo and Uber lawyers to clear out of his courtroom, the odds of doing that appeared to plummet.

After Anthony Levandowski refused to get a testing permit to run Uber's autonomous cars in California in December of 2016, he accepted Arizona governor Doug Ducey's invitation to test in his state. It was a good testing ground. The sunny weather was kind to sensors that weren't so good in snow and rain. Most streets, developed for a car-dominated transportation network in the second half of the twentieth century, were wide and straight. Few people walked or cycled anywhere, minimizing complicating factors. Best of all, the state put virtually no limits on who operated within its borders, or how. That was key, because every serious competitor believed that testing on public roads was the only way to properly train and evaluate the tech, and that paying a human to sit in the driver's seat and retake control if necessary was the way to keep everyone safe.

At about 9:00 the night of Sunday, March 18, 2018, Rafaela Vasquez climbed into one of the Volvo XC90 SUVs that Uber had outfitted with its suite of sensors and self-driving software. She had been given the route she was supposed to drive on a loop for her nearly eight-hour shift, noting any problems in the custom tablet that took the place of the car's center screen. Before she had started as a safety operator for Uber in the Phoenix suburb of Tempe, forty-four-year-old Vasquez had taken a three-week training course, including a week in Pittsburgh, with classroom time spent going over the technology and testing protocols, and time on a track learning how to maneuver a car out of dangerous situations. She'd been trained to keep her hands an inch or two from the steering wheel and her right foot hovering over the brake pedal. She'd been told to remain vigilant and be ready to take control of the vehicle, and that using one's phone while the car was driving was a fireable offense. Uber's tech had improved markedly in the past year, but it was nowhere near reliable enough to operate without human supervision.

At 9:58, the Volvo was driving north in the right lane of Mill Avenue, going the speed limit in a 45 mph zone. The car's radar, Lidar, and cameras detected the presence of a forty-nine-year-old woman named Elaine Herzberg, who stepped from the median into the road, pushing an orange and black bicycle loaded with plastic bags. As Herzberg walked across the shoulder and the two left lanes, Uber's software alternately classified her as a vehicle, a cyclist, and an unidentified object, an investigation by the National Transportation Safety Board later revealed. It did not identify her as a person on foot because Herzberg was jaywalking. Uber hadn't taught its cars to look for pedestrians outside of crosswalks, the safety investigators found. It was a galling failure of imagination and a sign that, even after years of work, the makers of robo-cars didn't always appreciate how their technology had to adapt to a human world.

When just twenty-five meters separated Herzberg and the car, the computer determined it needed to slam on the brakes. But Uber's engineers had limited the car's ability to make an emergency stopping maneuver, fearful of it making the wrong call and causing a crash by halting for no reason. That's why they had the safety driver there, after all. But dash cam footage released by the police showed Vasquez wasn't keeping her eyes on the road. She was looking down at something out of the camera's field of view, near her right knee. According to the National Transportation Safety Board, in the three minutes before Herzberg started her crossing, Vasquez took her eyes away from the road twenty-two times. Seven of those looks lasted more than three seconds.

Vasquez told investigators she was looking at the tablet that displayed information about the autonomous system, but a police inquiry determined that her silver LG smartphone was streaming an episode of the NBC singing competition *The Voice* at the time. Whatever the truth, she didn't see Herzberg until a fraction of a second before the car struck her at nearly 45 mph, throwing her 75 feet. Vasquez stopped the car and dialed 911. Herzberg died at the hospital that night, the first bystander killed by a self-driving car.

By Monday morning, the crash was national news. Uber immediately parked its cars in Tempe, Pittsburgh, San Francisco (where it had finally gotten a permit and done some testing), and Toronto, where it had hired a team of high-profile artificial intelligence researchers after firing Levandowski as its self-driving lead. Developers like Waymo were quick to argue that their system would not have made the same mistake. Uber was among the few companies that put just one safety operator in its cars, which made it easier for that person to break the rules. Worse, Uber insiders said, the cars had been performing terribly. One had even driven onto a Pittsburgh sidewalk. In a March 13 email, Robbie Miller, an Uber operations manager who'd helped develop Chauffeur's testing program, alerted his bosses to serious problems. "The cars are routinely in accidents resulting in damage. This is usually the result of poor behavior of the operator or the AV technology. A car was damaged nearly every other day in February. We shouldn't be hitting things every 15,000 miles," he wrote in an email later published by *The Information*. "Repeated infractions for poor driving rarely results [sic] in termination. Several of the drivers appear to not have been properly vetted or trained." But critics inside the company said that Uber's self-driving leadership prized logging miles as a metric to show investors and the public that the program was humming along in the wake of the Waymo lawsuit and Levandowski's departure. Miller's manager told him the company would look into his concerns. Uber's car killed Herzberg five days later.

Whatever Uber's faults may have been, the crash threatened to impugn the very idea of a self-driving car. This was the sort of death that robots should prevent. Uber's failure raised questions that the young self-driving industry had so far elided, like who was liable, financially or criminally, in the event of a crash. Uber quickly reached an undisclosed settlement with Herzberg's family. Authorities eventually charged Vasquez with negligent homicide in September 2020; she pleaded not guilty. Uber evaded prosecution, despite criticism that it had set her up to fail. But the tragedy left open the question of whether cities and states should be so eager to have robots roaming their streets, even if it gave them the gloss of being home to new technology. Arizona governor Doug Ducey—who had welcomed Uber "with open arms and wide open roads" in 2016—banned the company's robo-cars from the state. Uber soon shuttered its Tempe operation, firing all 254 safety operators who staffed it, including Vasquez.

The aftermath of the crash didn't answer the harder questions. What were the ethics of testing on public roads, around people who had no choice but to participate in a science experiment? Was having a human in the car, or even two, enough to guarantee the public's safety? And if the technology was still making such simple errors, with disastrous consequences, when might it be ready to start saving lives instead of taking them? How would its creators know when it was ready, and how would they prove it to a rightfully wary public?

By the end of 2018, that public would be asked to make a leap of faith.

Every year around late October or early November, the Parks Department workers of Chandler, Arizona, started gathering tumbleweeds. They roamed the outskirts of the Phoenix suburb, grabbing the prickly, uprooted, rolling bushes and tossing them into a custom-made trailer. When they had gathered a thousand or more of the things, looking for a variety of shapes and sizes, they attached them to a twenty-five-foot-tall, conical chicken-wire frame. After painting the result white, they covered it with flame-retardant chemicals, sixty-five pounds of glitter, and about twelve hundred holiday lights. In the Sonoran Desert, this was what passed for a Christmas tree.

Chandler is a predominantly white and wealthy city of 250,000 people, a land of large houses with big yards, palm trees, cacti, a skeletal public transit system, strip malls, and indoor malls. As 2018 drew to a close and municipal workers celebrated by lighting that year's tumbleweed tower, John Krafcik announced that Waymo was ready to deliver the future. And that Chandler, with its good weather and friendly regulatory environment, would be its first customer.

Waymo had been running a proto-ridehail service in the city for close to two years by that point. In April 2017, it had selected a few hundred people to participate. Using an app on their phones, they could call a Waymo and use it to get around town. They would sit in the second or third row of the white, Chrysler Pacifica minivans, more easily identified by the black, gumdrop-shaped sensor on the roof containing Waymo's proprietary Lidar, than by the green-and-blue "W" logo on the side. Unlike the adventurous Firefly pod car, these vehicles hadn't been deprived of their steering wheels or pedals. And good thing, because behind the wheel sat a Waymo safety operator, there to answer any questions and ensure that the vehicle stayed out of trouble.

This was where the team that had started life in 2009, as Google's Project Chauffeur, explored the logistical realities of operating about a hundred cars. Waymo set up a sixty-eight-thousand-square-foot depot and struck a deal with rental car company Avis to help to maintain and clean its fleet. It huddled with the local fire and police departments, participating in a test to prove its tech could detect all sorts of sirens and pull over safely. It set up a call center where agents could handle rider questions, and, more importantly, help the cars if necessary. This would prove to be a common yet little discussed feature of any robo-ridehail service, premised on the fact that no self-driving vehicle could ever be infallible. If a car without a driver inside got into a situation it couldn't handle—an unexpected construction zone maybe, or a cop directing traffic the wrong way down a one-way street—it would slow to a stop and send out a digital request for aid. From Waymo's remote centers, a worker would inspect the scene using the car's cameras, determine what to do, and issue the car instructions, along the lines of *Cross the double yellow, proceed ten meters, and return to the right side of the road.* The car, still driving itself and using its sensors to watch for trouble, would execute the maneuver before returning to its normal operating procedure.

Krafcik had promised a commercial service launch sometime in 2018, and so on December 5 of that year, Waymo announced "Waymo One," the real-deal version of its prototype service. Riders would no longer be bound by nondisclosure agreements that stopped them from discussing their experiences, and they would pay for rides at prices comparable to what Uber and Lyft charged for their human-driven services. They could call a car at any time, and go anywhere within a roughly one-hundred-square-mile area that included Chandler and the neighboring cities of Mesa, Gilbert, and Tempe. The minivans had room for three adults and one kid (the child's seat wasn't to be removed). Users could monitor their ride using one of the screens fixed to the back of the front row seat headrests. The display offered a dark-toned simulacrum of the road, depicting the Waymo in the middle and highlighting other vehicles, cyclists, pedestrians, and traffic signals the car detected. If that wasn't enough to assuage a nervous rider, it offered options to have the car pull over, or to dial up a customer service rep trained to answer questions. Otherwise, riders could relax and do whatever they liked, checking the car's route and ETA on the display.

Not everyone was enthusiastic about Waymo's presence. One person slashed a car's tires. One threatened a safety operator with a PVC pipe, another waved a revolver. On multiple occasions, a man driving a Jeep Wrangler tried to force a Waymo minivan off the road. His wife told the *New York Times* that she had done the same, explaining that one of the robots had nearly hit their ten-year-old son. Other, less confrontational Chandler residents simply found the Waymo cars annoyingly slow and cautious. Not that their grievances carried any weight: State and local authorities fully supported Waymo's operation in the area, saying the tech company brought jobs to the region.

The service that Waymo's press team pitched as an epochal change, though, came with two important caveats. First, it wasn't open to the general public. Only those riders preselected for the pilot could call up a robot, at least to start. Second—the one that really mattered—was that while the cars would drive themselves, they were not *driverless*. Waymo would continue to pay its safety operators to sit behind the wheel. The computer its engineers had spent a decade developing just wasn't ready to go without its human backup.

Yes, it had been almost exactly ten years since January 2009, when Sebastian Thrun, Chris Urmson, Anthony Levandowski, and their teammates embarked on a voyage to realize the vision of DARPA's Challenges. These engineers had put Waymo on a path to drive 10 million miles on public roads and billions more in computer simulations. They were the first members of a team that grew to include hundreds of people honing algorithms and building maps and tinkering with hardware, all to root out and answer every last *what if*. Of the original team, few had stuck around for the era of Waymo and John Krafcik. Most moved on to their own robo-ventures, each applying his knowledge to craft his own take on the effort, just as they had brought their own approaches to the Challenges. But none were surprised to see Waymo continue to rely on its humans. If they had realized anything in the decade since they'd first joined forces, it was that they were pursuing a problem that put a cruel spin on Sisyphus: The higher they pushed their boulder, the steeper the climb became, the more opaque the clouds that blocked their view of the summit.

Waymo called its car the "world's most experienced driver." The snappy marketing line ignored the difference between experience and wisdom. For all those miles, Waymo—and its now myriad competitors—still struggled to disprove Moravec's Paradox. With a few years of experience, any human driver could handle a car capably and in just about any conceivable situation. Making a robot do the same, reliably enough to underpin a real-life business, was almost impossibly harder. More remarkable than the stubbornness of the problem, though, was the stubbornness of those who had decided they would crack it—no matter how long it took or how much it cost.

The underwhelming launch of Waymo One was not a signal of failure. It wasn't the equivalent of Anthony Levandowski's motorcycle falling over, or Chris Urmson's Sandstorm burning up its tires on Daggett Ridge, or Sebastian Thrun's Junior losing out to the faster, more aggressive Boss. This race had no deadline, no first place. It was not the Grand Challenge but a grand challenge. And even if no one knew exactly where the finish line lay, or what reward waited on the other side, nothing would stop them driving toward it.

Epilogue: The Starting Line

EARLY IN THE MORNING OF A WARM, CLEAR FRIDAY IN OCTOBER of 2018, Anthony Levandowski sat down where everyone who knew him had predicted he would put himself before too long: behind the wheel of a Toyota Prius he had programmed to drive itself, ready to reclaim his place in the world he had helped create. As he headed over the Golden Gate Bridge, Levandowski could see the San Francisco skyline over his right shoulder. To his left was the open expanse of the Pacific Ocean. Ahead were the green-brown hills of Marin County, where he had started his life as an American entrepreneur after a childhood in Belgium. But Levandowski wasn't headed home. He was starting a nearly nonstop, three-day journey that would take him through Sacramento and Reno, past Utah's Bonneville Salt Flats, over Nebraska's plains and through Iowa's cornfields, dipping southeast through Indianapolis, swinging northeast past Pittsburgh and then straight east through New Jersey and over the George Washington Bridge. As the sun rose on Monday morning, the Prius crossed the Hudson River into New York City, completing the cross-country mission.

Levandowski had taken the wheel to make stops for gas and hotels, he said. Otherwise, the car had done all 3,099 miles of driving, handling a country's worth of changing types of roads and traffic, with a significant caveat: His latest creation was back where the Chauffeur team had started in 2009, on the highway. It stayed between the lane lines and a safe distance from other vehicles, occasionally changing lanes to get around slower (presumably human) drivers. The point of the coast-to-coast drive was to show that the system, while simple, was robust. Levandowski was so set on making it to New York without grabbing the wheel that he had turned back and started over twice, first in Utah when the system disengaged itself amid heavy winds on a banked curve, and then after being pulled over by a Nevada cop who was curious about the car going under the speed limit.

Two months after rolling into New York City, Levandowski made his return official. "I know what some of you might be thinking: 'He's back?' Yes, I'm back," he wrote in a blog post announcing the completion of the drive and the launch of his latest company, Pronto. He had put together this venture in January of 2018, even before the conclusion of *Waymo v. Uber.* "I'm back because it's my life's passion to make the life-saving potential of autonomous vehicles a reality." The post echoed, in nobler tones and better grammar, the emails he had sent to Google's Larry Page about his dissatisfaction with the Chauffeur project. "Over the past 15 years," it read, "we've witnessed numerous advances in self-driving technology. I'm proud to have played a big role in it. At the same time, I've admittedly grown frustrated—and at times impatient—with the industry's inability to deliver on its promises." Levandowski elided the black marks on his résumé, brazenly blaming his reputation for recklessness on his "openness to bluntly address technology's capabilities and its limitations." But he also made clear that the age of Pronto marked not just new letterhead, but a new approach. "The self-driving industry has gotten two key things wrong," he wrote. "It's been focused on achieving the dream of fully autonomous driving straight from manual vehicle operation, and it has chased this false dream with crutch technologies."

The first criticism referred to the "moonshot" approach that Chris Urmson and his teammates, including Levandowski, had settled on years earlier, the decision not to phase out the human driver, but to dump her altogether. Now Levandowski was quick to point out that whatever Waymo had accomplished in the Phoenix suburbs, it was no Apollo program. It still had humans behind the wheel. The better way to deliver on the technology that had dominated his life since he started making a self-riding motorcycle was in incremental steps, Levandowski had decided. Start with a highway system that, like those built by Tesla and Cadillac, required the human to pay attention to the road. Then build from there. He would sell his aftermarket system to long-haul truckers, a population that stood to benefit from tech that made driving easier and safer.

One of the "crutch technologies" he was talking about was the extremely high-resolution maps that nearly every robo-car on the road relied on. The challenge of creating and maintaining those records was knotty enough to produce its own mini-industry of companies roaming the roads with mapping vehicles. Cartography had been key to autonomy since before the era of the Grand Challenges. It was how Sebastian Thrun's early robots moved through museums in Germany and the United States. It took the pressure off a vehicle's ability to read things like stop signs and speed limits, but in exchange for a leash on where the vehicle could go.

The second crutch was a technology Levandowski had evangelized ever since he saw it spinning atop Dave Hall's pickup truck at the 2005 Grand Challenge. He had been the first real automotive Lidar salesman, pitching it to Urban Challenge teams around the country. He had used his company, 510 Systems, to develop his own version of the tech, then sold it to Google. He had pitched it as a way to accelerate Uber's development effort, and lived through a devastating legal fight over it. Now he said self-driving cars didn't need Lidar at all. It was too expensive, too limited in range and resolution, too unreliable to be used on a car. Levandowski was betting that with the right deep-learning software, he could teach a car to drive itself using nothing but images from cameras. If he could, he would dodge the constraints that threatened to limit the growth of programs run by the likes of Waymo: the need for detailed maps and many, many Lidar scanners, which no one had yet managed to make affordable or reliable. He would have a system that, like the most elegant and lucrative kind of software, scaled.

Anthony Levandowski was back, but he wasn't quite the same young man who'd helped launch Google's effort a decade earlier, or the graduate student whose robotic motorcycle ended up in the Smithsonian. He was worth $72 million, for one thing. But like your average nearly-forty-year-old, he wasn't as skinny as he used to be. His smartphone background featured a photo of his two young sons. He was less dictatorial, said his DARPA Grand Challenge teammate Ognen Stojanovski, who had joined him at Pronto, along with other old allies. Less arrogant, and more willing to hold off on making decisions, to sacrifice a bit of time in the interest of making the right call. When it came to putting self-driving tech to use, the moon was a shot too far, Levandowski said. "I'm trying to get onto the garage roof."

The climb would have to wait. On August 27, 2019, Levandowski was indicted on thirty-three federal charges of theft and attempted theft of trade secrets. Two and a half years after Waymo had filed its lawsuit against Uber, the Department of Justice was going after the man at the center of the case. Nobody had ever denied that Levandowski downloaded a trove of documents before leaving Google. Now his fate would hinge on whether prosecutors could prove that those documents included trade secrets, and that he had taken them in an effort to enrich himself. Each count could cost him a fine of $250,000 and a ten-year prison sentence. Levandowski pled not guilty.

At his second bail hearing, two weeks later, Levandowski towered over his two lawyers as he stood before the judge. He let his counsel do the talking, only speaking when spoken to, and then only to say, "Yes, Your Honor." He resigned as head of Pronto. To make up his $2 million bail, Levandowski's father and stepmother had put their house down as collateral. His longtime friend Randy Miller—who had sat with him in a hot tub sixteen years earlier thinking a self-driving motorcycle would be cool—had done the same. On one ankle, Levandowski wore an electronic monitoring device, extra insurance he wouldn't leave the country, or even go near an airport. On his feet, the engineer who always wore sneakers, refusing to sacrifice comfort for societal conventions, wore a pair of shiny black dress shoes. The ensuing indignities proved less symbolic.

In March of 2020, in a separate legal fight, an arbitration panel ruled that Levandowski had engaged in unfair competition when he left Waymo, and that he owed his former employer $179 million. Suddenly, $72 million didn't sound like so much. He filed for bankruptcy. And a few weeks later, he negotiated a deal with the DOJ, pleading guilty to one count of trade secret theft in exchange for the rest of the charges being dropped. In the agreement, he admitted that he downloaded the roughly 14,000 documents "with the intent to use them for my own personal benefit." He assented to allow any law enforcement officer "at any time, with or without suspicion," to search him, his home, car, office, and electronic devices for evidence of malfeasance. He agreed to pay Waymo $756,499.22 to cover what it had spent to help the government investigate the case.

"I was excited about fighting and winning," Levandowski said. "Having Google be scared of you is a pretty big compliment." But he had realized this case wasn't worth fighting. "I'm happy to put this behind me."

Levandowski had always rushed ahead, his eyes straining for a look around the next corner, his long legs whisking him along. But when he reported to court for his sentencing in early August, he came in a reflective mood. "The last three and a half years have forced me to come to terms with what I did," he told Judge William Alsup, who had also tried the *Waymo v. Uber* case and recommended the Department of Justice investigate Levandowski. The engineer apologized to his former colleagues; he thanked his family members and friends who had stood by him. "I will never come close to breaking the law again," he said.

Alsup seemed to sympathize, and called Levandowski a great engineer. "I respect that. I want you to know that," he said. But he demanded more than contrition. "This was not a small crime," the judge said. He wanted to deter others from stealing trade secrets, to set an example. He ordered Levandowski to pay a fine of $95,000, and, starting after the COVID-19 pandemic had passed, to serve eighteen months in prison. After his release, Alsup ruled, Levandowski must give a speech (or speeches) to at least two hundred people, to be titled "Why I Went to Federal Prison."

After more than a decade of research into self-driving technology, Sebastian Thrun's admonition to the Chauffeur team years earlier— "We have not yet saved a single life. We have not yet enabled a single blind or disabled person to operate a car"—was as true as ever. Instead, Elaine Herzberg was dead, run down in Tempe, Arizona, by the robot that was supposed to keep her safe.

Where autonomy wasn't dangerous, it felt disappointing. Waymo's competitors liked to snigger about its inability to take the human out of the driver's seat in its Arizona robo-taxi service, but no one else had gone any further. General Motors' Cruise announced that it would miss its self-imposed deadline to launch its service by the end of 2019, and didn't offer a revised timeline. Ford continued to target 2021 for putting self-driving cars on the road en masse, but Argo's Bryan Salesky made clear that his outfit wouldn't deploy anything before it was ready. Uber's program limped along, trying to recover from the body blow it had dealt itself just weeks after dodging Waymo's knockout punch. Elon Musk kept promising that a fully self-driving Tesla was just around the corner, and kept failing to deliver.

All the while, Americans were still, on average, wasting an entire week in traffic every year, costing the economy more than $300 billion. After decades of decline, deaths on the road had spiked to nearly forty thousand a year. One of the many lives claimed along the way was that of Brian Lynn, whose pickup truck was hit head on by a Toyota sedan that drifted across the center lane of California Highway 247. The local newspaper noted that Lynn owned Barstow's Slash X Cafe, but didn't connect that "shit-kicking cowboy-type place out in the middle of nowhere" to the 2004 DARPA Grand Challenge it had hosted, or to its role in boosting the technology that would come too late to save him.

By the time of Levandowski's criminal case, media coverage of self-driving cars had became more skeptical, TED talks less frequent. Sebastian Thrun wasn't the only person saying that the *flying cars* he was building for Google founder Larry Page would have a serious impact on human transportation before self-driving did. These electric aircraft would carry a few people on short hops around metro areas, leaving the landlubbers to suffer in traffic. Battery technology and regulatory hurdles were a challenge, but getting something to fly itself was easy. Anyone who knew anything about unmanned technology knew that.

To those who followed the history of technology, that change didn't mark the death of the self-driving dream, but the "trough of disillusion-ment." This was the third stage on the "hype cycle" that research firm Gartner created to explain the development of new technologies, fol-lowing the "innovation trigger" and the "peak of inflated expectations." The trough was where public interest waned, where reality hit home. The only way out was the "slope of enlightenment," where the company or industry backed off the promises and homed in on the tech's specific benefits, where new iterations of inventions came to bear on the prob-lems that had beguiled their predecessors. Not every technology made that climb, but those that did were bound for the "plateau of productiv-ity." By 2020, self-driving looked to be on the way up the slope.

An effort that started in the Mojave Desert was now a blooming in-dustry with hundreds of participants, each targeting some niche with its own blend of technology and business plan to get there. Waymo, of course, was taking the big swing by aiming for a fully driverless robo-taxi system that it would bring to one city after another. So was GM's Cruise, which was now led by Dan Ammann. Ammann had resigned as president of the automaker to run the outfit, and believed that might made right. When GM acquired Cruise, the startup had forty employ-ees. By 2020, it had eighteen hundred. "I can't think of another example of a situation where you have to get to the point where you have literally thousands of engineers, billions of dollars of capital, and no product yet," Amman said. "This is not a problem you can solve with fifty or a hundred engineers."

Those smaller companies took a different tack. Don Burnette had spent years with Google, watching the team balloon in size and log millions of miles on the road. Every mile cost money and came with the risk of a crash, and so Burnette kept his small, experienced Kodiak Robotics team focused on real-world driving that would teach his robots new, important lessons. "I want to succeed with as few miles as possible," he said. This idea of quality over quantity had become popular, as companies like Chris Urmson's Aurora preached similar values.

Kodiak was one of at least a dozen companies working on trucking, all lured by the economic upside of making robots work in a relatively simple highway environment. Others went the opposite direction, developing shuttles for carefully constrained areas—an echo of Chauffeur's experimental golf carts that once explored Google's campus. Those were just the folks making actual self-driving vehicles. A host of other players had emerged to help them along. These specialists did things like create maps, develop new Lidar sensors, and build systems to make remote operation centers more reliable and secure.

Of all these companies, many—perhaps most—would eventually fail. Fifteen years of work since the first DARPA Grand Challenge had shown that there was nothing easy about this. To make this Hobbesian view of life even bleaker, a career spent working on autonomous tech promised to be nasty, brutish, and long. Such was the price of climbing the slope of enlightenment. But surely some would make the journey. The question was no longer *if* self-driving vehicles would arrive, but *where*, and *in what guise.* In that way, the advent of the self-driving car looked to mirror the spread of the internet, DARPA's great accomplishment of the 20th century. It wouldn't hit everyone, or everywhere, all at once. It would arrive in pockets, for specific use cases in specific places, just as the internet first hit universities and government agencies. Autonomous tech might start with taxis in some downtown areas, or grocery deliveries in some suburbs. But as its purveyors continue to master its technological and logistical challenges, they will push it to more people and find new ways to propagate (and monetize) its benefits. Over the course of decades, their work will influence the many ways in which people move around today. And like access to the online world, it will become so pervasive that its absence will be hard to imagine. But such shifts are rarely as positive as their champions predict, or hope.

In 1939, as the world was just beginning its descent into a vicious conflict, millions of Americans lined up for the chance to catch a glimpse of the transportation future. When the United States came out the other side of the war as a bona fide superpower, industrial giants like Ford and General Motors, whose might had helped it secure that victory, turned their energies to making real the blockbuster attraction of the New York World's Fair. Sponsored by GM, Futurama had shown its visitors a world where infrastructure that catered to personal cars would produce a particular kind of freedom, the freedom of movement, delivered by prosperity. America—along with countries around the world—spent the next fifty years following that blueprint, catering to cars and constructing national networks of interstates, only to cement itself into a system that turned both highways and precious urban space into parking lots.

The second decade of the new millennium offered a chance to slide some TNT into all that concrete. Self-driving technology could make vehicles not just less deadly, but more efficient. But that was just one piece of a larger upheaval. Against this backdrop, GM CEO Mary Barra's prediction that the car business would change more in the next five years than it had in the past fifty looked small in scope. It wasn't just driving that was shifting. It was the fundamental notion of how to get from A to B.

A new idea of urbanism was taking root, fueled by grassroots, official, and corporate efforts. More than one thousand cities welcomed bike-share programs, and competed to create networks of bike lanes. Opening a new bit of public transit or turning a car-crowded venue like New York City's Times Square into a pedestrian haven were guaranteed ways for politicians to embellish their credentials. Uber, Lyft, and their ilk continued to make it easier to get around without owning a car, and even moved into the greater transportation field by launching fleets of shared bikes and electric scooters. Investors put millions of dollars into companies working on "hyperloops," essentially high-speed trains in near-vacuum tubes. The idea of flying cars became dizzyingly serious, with dozens of efforts, from the likes of Airbus and Boeing, trying to make electric-powered, vertical takeoff and landing vehicles to ferry passengers through the air and over traffic. Elon Musk found time between running Tesla and his rocket company SpaceX to propose digging massive networks of underground tunnels. He even bought a tunneling machine and created a test track under Los Angeles.

Watching people queue up to take in the 1939 Futurama exhibit he had designed, Norman Bel Geddes offered a confident diagnosis: "All these thousands of people who stood in line ride in motor cars and therefore are harassed by the daily task of getting from one place to another, by the nuisances of intersectional jams, narrow, congested bottlenecks, dangerous night driving, annoying policemen's whistles, honking horns, blinking traffic lights, confusing highway signs, and irritating traffic regulations; they are appalled by the daily toll of highway accidents and deaths; and they are eager to find a sensible way out of this planless, suicidal mess. The Futurama gave them a dramatic and graphic solution to a problem which they all faced."

The interstates put forward by Bel Geddes improved driving with innovations like cloverleaf intersections and wider lanes. But they also made it easy, in many places necessary, to rely on the personal car as means of transport. Generations of city planners prioritized moving vehicles, not people. Funding for public transport eroded, the idea of biking as an activity fit for those old enough to get a driver's license appeared silly.

Eighty years later, autonomous technology promised a similar pain relief, a new way to make life in the car more logical and less dangerous. It was a fantastic thing, a remarkable accomplishment, about as exciting a technological marvel as one could hope for. But as a tool, its success would be measured not by how many human drivers it took off the roads, but by how much it improved their lives. History was a cautionary tale, but the likes of Anthony Levandowski and the denizens of Silicon Valley had never kowtowed to the past.

When Tony Tether created the original DARPA Grand Challenge, he thought he was looking for the secret sauce, the way to combine existing hardware and software into a vehicle that could wander the Earth on its own. He was right, but not quite in the way he imagined. Fifteen years later, so many of the young people who'd been in the desert to watch a robotic massacre were still fighting the technology and one another. The very notion of what they were looking for had changed, from the just right combination that would solve the problem, to an acceptable one that would solve *a* problem, somewhere, in some context.

Success remained elusive. Machine learning wasn't the sauce. Mapping wasn't the sauce. Laser wasn't the sauce. But Tether had found what he was looking for. The sauce was the very fact of attracting all those people to the problem in the first place. Bringing them together and pointing them in the same direction. First in the desert, then in the city. Then at Google, where they defied conventional wisdom and ignored shortsighted investors to launch an effort that might finish what the Challenges had started. Then at Uber, Ford, General Motors, and so many more.

The ignominious end of the 2004 Grand Challenge was just the start of a much greater race. No one has reached the finish line yet; no one is sure where it lies, exactly. But many are racing toward it. Someday, somehow, some of them will get there.

Acknowledgments

First off, I want to thank everyone who didn't just live this story, but took time out of busy, robot-forging lives to tell me all about it, and to share the primary documents—photos, emails, letters to the US Fish and Wildlife Service, and more—that restored color to sepia memories.

All that would have meant little, though, if not for many others who made me just enough of a reporter and writer to weave together their tales. Of the many stories I wrote for *WIRED* magazine, few would have been much good if not for the sharp pencils and blunt criticisms of Joe Brown, Chuck Squatriglia, Adam Rogers, Andrea Valdez, Sarah Fallon, and Scott Thurm. They and others have made me the journalist I am, and I shall remain in their debt. Aarian Marshall, another terrific person I'm glad to call a friend as well as a colleague, not only helmed *WIRED*'s transportation desk while I took book leave, but offered invaluable insight and guidance on early drafts of this project.

Rich Dorment guided me through the oral history project on the first DARPA Grand Challenge that made me realize just how much of this history sat untold. And when I couldn't stop retelling the stories I'd heard in the process, my good friend Nick Stockton responded with a thought that hadn't occurred to me: "You should write a book." Amid bouts of stress and doubt, Sarah Frier and Tim Higgins, friends writing their own books, provided encouragement and the vital reminder that I wasn't alone. Ryan Loughlin was my coproducer of a mini-documentary on the three DARPA Challenges, and a butt-saving archivist of interviews I had feared lost. The New York Mets kept their 2019 season alive just long enough to provide a distraction when I needed it. Here's to next year, as always.

My agent, Eric Lupfer, helped shape a rambling sentence of an idea into something somebody might call a book proposal, and offered crucial advice on a years-long journey. Jacob Shea did the vital, often thankless work of fact checking this book. I apologize to him for my not quite perfect record keeping. I can't imagine a more helpful editor than Bob Bender for this, my first book. I appreciate especially his guidance on how to triage my many concerns, from valid (some) to insignificant (most). Thanks also to Johanna Li and the rest of the team at Simon & Schuster for their help getting this book from my brain to readers' hands.

Tim Wyman-McCarthy couldn't care less about cars or robots, but he did more than let me prattle on about this book, even while I flossed. He listened, offered steady encouragement and critiques, and displayed a talent for asking just the right question: the one that provokes thoughtful silence. My love and thanks to him, even if he did it just for a bit of quiet.

My parents, brothers, and the rest of my family have spent the most time of anyone dealing with and shaping me. With no real way to pay back that sort of debt, I'll settle for a simple thank-you.

Notes

Epigraph

vii *They have no fear of damages or loss* Homer, *The Odyssey,* translated by Emily R. Wilson (New York: W.W. Norton & Company, 2018).

Prologue

1 *"Where do you work currently?"* April 14, 2017, deposition of Anthony Levandowski, *Waymo LLC v. Uber Technologies, Inc.; Ottomotto LLC; Otto Trucking LLC* (United States District Court Northern District of California). For subsequent Levandowski testimony references, see same filing.

5 *hit the road annually* Nikolaus Lang, Michael Rüßmann, Jeffrey Chua, and Xanthi Doubara, "Making Autonomous Vehicles a Reality: Lessons from Boston and Beyond," Boston Consulting Group, October 17, 2017, https://www.bcg.com/publications/2017/automotive-making-autonomous-vehicles-a-reality.aspx.

5 *way more relaxing to be stuck in traffic* Roger Lanctot, "Intel Predicts Autonomous Driving Will Spur New 'Passenger Economy' Worth $7 Trillion," *Strategy Analytics,* June 2017, https://newsroom.intel.com/news-releases/intel-predicts-autonomous-driving-will-spur-new-passenger-economy-worth-7-trillion/#gs.o1eivj.

Chapter 1

9 *would be transferred to museums* H.R. 5408, 106th Cong., U.S. G.P.O. (2000) (enacted).

9 *nothing to say about Section 220* United States Government Publishing Office, *Weekly Compilation of Presidential Documents* 36, no. 44: 2690–93.

10 *who helmed the Armed Services Committee* US Congress, Congressional Record—Senate, 106th Cong., 2nd sess., July 14, 2000.

10 *might change that attitude* John Warner, phone interview by author, San Francisco, January 17, 2018.

10 *"We certainly wanted to challenge them."* Les Brownlee, phone interview by author, New York, December 27, 2017.

12 *put back in 1996* Annie Jacobsen, *The Pentagon's Brain: An Uncensored History of DARPA, America's Top Secret Military Research Agency* (New York: Back Bay Books/ Little, Brown and Company, 2016); Sharon Weinberger, *The Imagineers of War: The Untold History of DARPA, the Pentagon Agency That Changed the World* (New York: Alfred A. Knopf, 2017). For subsequent DARPA history references, see same.

15 *"the desire to be a science fiction writer"* "Dr. Tony Tether 2001–2009," interview, https://www.esd.whs.mil/Portals/54/Documents/FOID/ReadingRoom/DARPA /15-F-0751_DARPA_Director_Tony_Tether.pdf.

15 *budget increased 50 percent, to $3 billion* Defense Advanced Research Projects Agency, *Department of Defense FY 2002 Amended Budget Submission June 2001*, https://upload.wikimedia.org/wikipedia/commons/4/48/Fiscal_Year_2002 _DARPA_budget.pdf; Defense Advanced Research Projects Agency, *Department of Defense Fiscal Year (FY) 2005 Budget Estimates February 2004*, https://www .darpa.mil/attachments/(2G12)%20Global%20Nav%20-%20About%20Us%20 -%20Budget%20-%20Budget%20Entries%20-%20FY2005%20(Approved).pdf.

16 *controlled by radio waves sent from the (human-driven) car behind it* " 'Phantom Auto' Will Tour City," *Milwaukee Sentinel*, December 8, 1926.

16 *going a set speed and staying a safe distance apart* B. Alexandra Szerlip, *The Man Who Designed the Future: Norman Bel Geddes and the Invention of Twentieth-Century America* (New York: Melville House, 2017).

16 *picking up electronic signals from the highway* "1958 Firebird III," GM Heritage Center Collection, https://www.gmheritagecenter.com/gm-vehicle-collection /1958_Firebird_III.html. a

17 *built by the Artificial Intelligence Center at Stanford Research Institute* "Shakey," Artificial Intelligence Center, http://www.ai.sri.com/shakey/, accessed July 16, 2019.

17 *each using the platform for his own application* Lester Earnest, "Stanford Cart: How a Moon Rover Project Was Blocked by a Politician but Got Kicked by Football into a Self-Driving Vehicle," Stanford University, March 11, 2018, https://web.stanford.edu/~learnest/sail/cart.html, accessed July 16, 2019.

18 *Moravec wrote in his 1980 dissertation* Hans P. Moravec, "Obstacle Avoidance and Navigation in the Real World by a Seeing Robot Rover," PhD dissertation, Stanford University, 1980.

19 *even changing lanes and overtaking other cars* "The PROMETHEUS Project
    Launched in 1986: Pioneering Autonomous Driving," Daimler Global Media
    Site, September 20, 2016, https://media.daimler.com/marsMediaSite/en/
    instance/ko/The-PROMETHEUS-project-launched-in-1986-Pioneering
    -autonomous-driving.xhtml?oid=13744534, accessed July 16, 2019.

20 *"That's a hell of a challenge," he thought* Scott Fish, phone interview by author,
    Berkeley, California, January 13, 2018.

21 *"an ecosystem of doing business in an innovative way"* Rick Dunn, phone interview
    by author, Berkeley, California, January 16, 2018.

22 *left Fairfax Avenue as a DARPA ally* "Dr. Tony Tether 2001–2009," interview,
    https://www.esd.whs.mil/Portals/54/Documents/FOID/Reading Room/
    DARPA/15-F-0751_DARPA_Director_Tony_Tether.pdf.

22 *"Disneyland is a land of dreams and fantasy becoming reality"* "DARPATech 2002
    Symposium," DARPATech 2002 Presentations, https://archive.darpa.mil/
    DARPATech2002/presentation.html, accessed July 16, 2019.

23 *"I knew it was a show"* Tony Tether, in-person interview by author, Pittsburgh,
    Pennsylvania, October 14, 2017.

23 *"We honestly did not know what we were doing"* Tony Tether, phone interview by
    author, San Francisco, May 17, 2017.

25 *a violation of the Endangered Species Act* "About the Endangered Species Act,"
    Florida Museum, April 4, 2018, https://www.floridamuseum.ufl.edu/discover
    -fish/general/about-the-endangered-species-act/, accessed July 16, 2019.

26 *But he was intrigued* Sal Fish, phone interview by author, San Francisco, May 18, 2017.

26 *to the local homeless population* Tony Tether, phone interview by author, San
    Francisco, May 17, 2017.

26 *We really might have something here* Ibid.

Chapter 2

28 *"I don't want to be someone just providing a commodity at a low price"* Bonnie
    Azab Powell, "At 22, Anthony Levandowski Is Already a Veteran Businessman,"
    UC Berkeley News, February 13, 2003, https://www.berkeley.edu/news/media
    /releases/2003/02/13_levandowski.shtml, accessed July 16, 2019.

28 *"They made computers real"* Anthony Levandowski, phone interview by author,
    Berkeley, California, March 4, 2018.

28 *portable electronic blueprint displays for use on construction sites* Bonnie Azab
    Powell, "At 22, Anthony Levandowski."

29 *"And we are going to figure that out together"* "Events: Conferences Archive,"

Grand Challenge Schedule/Events, February 22, 2003, https://archive.darpa
.mil/grandchallenge04/conference_la.htm. For subsequent references to conference proceedings, see same. a

31 *a few hours east of Pittsburgh* Michael Arndt, " 'Red' Whittaker: A Man and His
Robots," *Bloomberg Businessweek*, June 25, 2006, https://www.bloomberg.com
/news/articles/2006-06-25/red-whittaker-a-man-and-his-robots. a

32 *provide the space for him to land his hammer* "Robotics History: Narratives and
Networks Oral Histories: Red Whittaker," interview by Peter Asaro and Selma
Šabanovic, IEEEtv, April 17, 2015, https://ieeetv.ieee.org/history/robotics
-history-narratives-and-networks-oral-histories-red-whittaker. a

33 *"He gets how things work"* David Wettergreen, phone interview by author, Berkeley, California, August 4, 2018.

34 *"Some people think about stuff like that"* William Whittaker, phone interview by
author, San Francisco, May 13, 2017.

35 *"Life isn't 'Mother, may I' "* Ibid.

35 *"then providence moves too"* William Red Whittaker, "Red's Race Log," *Red Team
Racing* (blog), March 14, 2003, https://www.cs.cmu.edu/~red/Red/racelogs
.html#, accessed July 15, 2019. For subsequent references to Whittaker's race
journal, see same.

36 *He headed to America* Chris Urmson, phone interview by author, Pelham, New
York, June 5, 2017.

38 *That, at least, Ganassi could fix* Chip Ganassi, phone interview by author, San
Francisco, May 2, 2018.

41 *"the top was missing"* Kevin Peterson, in-person interview by author, San Francisco, February 15, 2018.

41 *"what we had in mind"* Chris Urmson, in-person interview by author, Palo Alto,
January 28, 2019.

42 *"I'd toss them off the team"* William Whittaker, phone interview by author, Pittsburgh, Pennsylvania, October 15, 2017.

43 *another undergrad-turned-zealot* Matthew Johnson-Roberson, phone interview
by author, Berkeley, California, February 22, 2018.

44 *crashed through the fence anyway* W. Wayt Gibbs, "A New Race of Robots," *Scientific American*, March 2004.

Chapter 3

45 *designates friendly forces in war games* Burkhard Bilger, "Auto Correct," *New
Yorker*, November 15, 2013.

48 *kookier foil to Red Whittaker's near-professional effort* Douglas McCray, "The Great Robot Race," *WIRED*, March 1, 2004, 132–39, and Ashlee Vance, "Robotic Road Trip on a Military Mission," *New York Times*, October 9, 2003.

50 *Charlie Smart said* Bilger, "Auto Correct."

50 *"think of what you'll tell your grandkids" vein* Ognen Stojanovski, in-person interview by author, San Francisco, May 29, 2018.

52 *"spinning wheels looked like snowblowers"* Whittaker, "Red's Race Log," January 29, 2004, accessed July 15, 2019. For subsequent references to Whittaker's race journal, see same.

52 *"Red loves groundhogs"* Kevin Peterson, in-person interview by author, San Francisco, February 15, 2018.

53 *"You just worked until you passed out"* Matthew Johnson-Roberson, phone interview by author, Berkeley, California, February 22, 2018.

56 *"That's fucked" The Million Dollar Challenge*, The History Channel, 2004, DVD.

56 *"we rolled the vehicle"* Phil Koon, phone interview by author, Berkeley, California, June 8, 2018.

Chapter 4

59 *would net him the reward* Dave Wootton, *Galileo: Watcher of the Skies* (New Haven, CT: Yale University Press, 2013); J. L. Heilbron, *Galileo* (New York: Oxford University Press, 2010). For subsequent references to Galileo, see same.

60 *on the maps they were filling in as they went along* Dava Sobel, *Longitude: The True Story of a Lone Genius Who Solved the Greatest Scientific Problem of His Time* (New York: Walker, 1995).

60 *dollars, ducats, francs, rubles, lire, pounds, and more Selected Innovation Prizes and Reward Programs*, March 20, 2008, https://keionline.org/misc-docs/research _notes/kei_rn_2008_1.pdf.

61 *driving the vehicle he had built with his brother Charles* Kat Eschner, "The Forgotten Car That Won America's First Auto Race," *Smithsonian*, November 28, 2016, https://www.smithsonianmag.com/smart-news/model-t-came-duryea -wagon-180961218/.

62 *"save the lives of untold numbers of US soldiers"* Tony Tether, "Welcoming Speech," DARPATech 2004, Anaheim, California, https://web.archive.org/web /20040619233454/https://www.darpa.mil/DARPAtech2004/pdf/scripts/Tether Script.pdf.

63 *site visits that covered both coasts* Doug Gage, phone interview by author, Berkeley, California, June 21, 2018.

63 *have their discussion at the diner down the road* Tony Tether, in-person interview by author, Pittsburgh, Pennsylvania, October 14, 2017.

67 *"Is anybody going to move today?"* Jose Negron, phone interview by author, Berkeley, California, July 20, 2018.

68 *leaned over to kiss his wife The Million Dollar Challenge*, The History Channel, 2004, DVD.

69 *but had performed well enough* Defense Advanced Research Projects Agency, "Fifteen Teams Selected to Participate in the DARPA Grand Challenge Field Test," news release, March 12, 2004, https://archive.darpa.mil/grandchallenge04/media/qid_results5.pdf, accessed July 18, 2019.

69 *"it merited moving forward, to at least try"* Jose Negron, phone interview by author, Berkeley, California, July 20, 2018.

70 *and not much else* "City of Barstow," History | City of Barstow, http://www.barstowca.org/visitors/about-barstow/history. a

70 *DARPA had recruited him to design* Sal Fish, phone interview by author, San Francisco, May 18, 2017.

74 *Now they had the carnage to match. Grand Challenge 2004 Final Report*, Defense Advanced Research Projects Agency, 2004.

74 *met the same fate a few seconds later* Chris Urmson, Joshua Anhalt, Michael Clark, Tugrul Galatali, Juan P. Gonzalez, Jay Gowdy, Alexander Gutierrez, Sam Harbaugh, Matthew Johnson-Roberson, Hiroki Kato, Phillip Koon, Kevin Peterson, Bryon Smith, Spencer Spiker, Erick Tryzelaar, and William "Red" Whittaker, "High Speed Navigation of Unrehearsed Terrain: Red Team Technology for Grand Challenge 2004," June 1, 2004, https://ri.cmu.edu/pub_files/pub4/urmson_christopher_2004_1/urmson_christopher_2004_1.pdf. For subsequent references to Sandstorm's behavior in the 2004 Grand Challenge, see same.

76 At least we get to go home now Matthew Johnson-Roberson, phone interview by author, Berkeley, California, February 22, 2018.

77 *"caught fire, blah blah blah"* Tony Tether, phone interview by author, San Francisco, May 17, 2017.

77 WIRED *sassed* WIRED Staff, "Foiled: Darpa Bots All Fall Down," *WIRED*, March 13, 2004, https://www.wired.com/2004/03/foiled-darpa-bots-all-fall-down/. a

77 *"Nobody even came close," CNN wrote* Marsha Walton, "Robots Fail to Complete Grand Challenge," CNN, May 6, 2004, http://www.cnn.com/2004/TECH/ptech/03/14/darpa.race/, accessed July 19, 2019.

77 *said tech site the Register* Ashlee Vance, "DARPA's Grand Challenge Proves to Be Too Grand," the Register, March 13, 2004, https://www.theregister.co.uk/2004/03/13/darpas_grand_challenge_proves/. a

77 *"continued to predict a victor"* Joseph Hooper, "From Darpa Grand Challenge 2004: DARPA's Debacle in the Desert," *Popular Science*, June 4, 2004, https://www.popsci.com/scitech/article/2004-06/darpa-grand-challenge-2004darpas-debacle-desert/. a

77 *"And this time, the prize will be $2 million."* Tony Tether, phone interview by author, San Francisco, May 17, 2017.

Chapter 5

79 *to keep them from his island* Adrienne Mayor, *Gods and Robots: Myths, Machines, and Ancient Dreams of Technology* (Princeton and Oxford: Princeton University Press, 2018).

79 *could sort a pile of blocks by size* Lester Earnest, "Stanford Artificial Intelligence Laboratory (SAIL)," Stanford University, https://web.stanford.edu/~learnest/sail/, accessed July 19, 2019.

80 *supporting two-thirds of the lab's 128 scientists* Lester Earnest, ed., *The First Ten Years of Artificial Intelligence Research at Stanford*, report no. STAN-CS-74-409, Computer Science, Stanford Artificial Intelligence Laboratory, July 1973.

80 *"require it to cross rooms many times"* Hans Moravec, *Mind Children: The Future of Robot and Human Intelligence* (Cambridge, MA: Harvard University Press, 1988). For subsequent references to Moravec's research and Moravec's Paradox, see same.

80 *in just nineteen moves* "Deep Blue," IBM100—Deep Blue, https://www.ibm.com/ibm/history/ibm100/us/en/icons/deepblue/. a

82 *gave short tours to more than two thousand visitors* Wolfram Burgard, Armin B. Cremers, Dieter Fox, Dirk Hähnel, Gerhard Lakemeyer, Dirk Schulz, Walter Steiner, and Sebastian Thrun, "Experiences with an Interactive Museum Tour-Guide Robot," *Artificial Intelligence* 114, no. 1–2 (October 1999): 3–55.

82 *how often those are cited by others* "Sebastian Thrun," Google Scholar, https://scholar.google.com/citations?user=7K34d7cAyAAAJ&hl=en. a

83 *invention of new techniques in the eighties* Larry Hardesty, "Explained: Neural Networks," *MIT News*, April 14, 2017, http://news.mit.edu/2017/explained-neural-networks-deep-learning-0414. a

83 *" 'My god, let's hope that's not all there is to it.' "* Sebastian Thrun, in-person interview by author, Mountain View, California, May 24, 2017.

85 *decided to start by building a prototype* Sebastian Thrun, "A Personal Account of the Development of Stanley, the Robot That Won the DARPA Grand Challenge," *AI Magazine*, December 15, 2006, 69–82. For subsequent references to the development of Stanley, see same.

 91 *landing in a thornbush* John Markoff, "Robotic Vehicles Race, but Innovation Wins," *New York Times*, September 14, 2005, https://www.nytimes.com/2005 /09/14/business/robotic-vehicles-race-but-innovation-wins.html, accessed July 18, 2019.

 93 *the old marine would bark at his recruits The Great Robot Race*, directed by Joseph Seamans, accessed July 18, 2019, https://www.pbs.org/wgbh/nova/darpa/ credits.html. For subsequent references to the Red Team's 2005 Grand Challenge effort, see same.

 95 *"I could do a hell of a lot better"* Jim McBride, phone interview by author, San Francisco, May 31, 2017.

 95 *"or drive off the edge of a cliff"* Ibid.

 98 *"I was a rug merchant"* Dave Hall, in-person interview by author, Alameda, California, November 18, 2018.

 98 *"it's got to be something other than that"* Dave Hall, phone interview by author, San Francisco, May 24, 2017.

 99 *map the surface of the Moon* "The Apollo 15 Lunar Laser Ranging RetroReflector," NASA: Apollo Revisited, April 2, 2019, https://www.nasa.gov/mission _pages/LRO/multimedia/lroimages/lroc-20100413-apollo15-LRRR.html, accessed July 19, 2019.

101 *bootlegger known as Whiskey Pete* Joe Oesterle, Tim Cridland, Mark Moran, and Mark Sceurman, *Weird Las Vegas: Your Alternative Travel Guide to Sin City and the Silver State* (New York: Sterling Publishing, 2007).

102 *wearing a bicycle helmet* John Markoff, "In a Grueling Desert Race, a Winner, but Not a Driver," *New York Times*, October 9, 2005.

103 *seven would reach the halfway point Report to Congress: Darpa Prize Authority*, Defense Advanced Research Projects Agency, 2006, https://www.grandchal lenge.org/grandchallenge/docs/Grand_Challenge_2005_Report_to_Con gress.pdf.

104 *Highlander's stumble wasn't a fluke* "Live Reports from the 2005 Darpa Grand Challenge," *Popular Science*, October 8, 2005, https://www.popsci.com/sci- tech/article/2005-10/live-reports-2005-darpa-grand-challenge/, accessed July 18, 2019.

106 *"We made it"* "Dr. Tony Tether 2001–2009," interview, https://www.esd.whs.mil /Portals/54/Documents/FOID/Reading Room/DARPA/15-F-0751_DARPA _Director_Tony_Tether.pdf.

106 *"This is the biggest day of my life"* Sebastian Thrun, in-person interview by author, Mountain View, California, September 18, 2017.

Chapter 6

107 *had dreaded a repeat performance* "Dr. Tony Tether 2001–2009," interview, https://
www.esd.whs.mil/Portals/54/Documents/FOID/Reading Room/DARPA/
15-F-0751_DARPA_Director_Tony_Tether.pdf. For subsequent references to
Tether and Total Information Awareness, see same.

107 *"seriously contemplated in the United States"* William Safire, "You Are a Suspect,"
New York Times, November 14, 2002, https://www.nytimes.com/2002/11/14
/opinion/you-are-a-suspect.html. a

108 *trumpeted the* New York Times Markoff, "In a Grueling Desert Race."

108 *said the* Washington Post Alicia Chang, "A Big Finish with No One at the Wheel."
Washington Post, October 9, 2005, https://www.washingtonpost.com/archive
/politics/2005/10/09/a-big-finish-with-no-one-at-the-wheel/7328a40f-44a4
-4545-8d21-0d09d3430fe6/?utm_term=.d9018130b60c. a

108 *one more driverless vehicle competition.* See: Defense Advanced Research Projects
Agency. "DARPA Announces Third Grand Challenge: Urban Challenge Moves
to the City." News release, May 1, 2006. Accessed July 20, 2019. https://archive
.darpa.mil/grandchallenge/docs/PR_UC_Announce_Update_12_06.pdf.

109 *surveying eighteen hundred acres of dry lakebed* R. J. Bowers and Eric Close, "A
Practical and Autonomous Geophysical Platform," *Symposium on the Application of Geophysics to Engineering and Environmental Problems*, January 2007,
749–54.

110 *"We've got to be involved"* Lawrence Burns, phone interview by author, Berkeley,
California, January 9, 2019.

110 *GM at the peak of its prowess* General Motors. "GM Marks Tech Center's National Historic Landmark," news release, August 6, 2015, https://media.gm
.com/media/us/en/gm/news.detail.html/content/Pages/news/us/en/2015
/aug/0806-lankmark.html. a

111 *DARPA no longer needed anyone and everyone.* Defense Advanced Research
Projects Agency, "DARPA Announces Third Grand Challenge: Urban Challenge Moves to the City," news release, May 1, 2006, https://archive.darpa.mil
/grandchallenge/docs/PR_UC_Announce_Update_12_06.pdf. a

111 *the involvement of some of its best researchers* Carnegie Mellon University, "Carnegie Mellon, General Motors Will Compete in 2007 DARPA Urban Challenge," news release, August 24, 2006, http://www.tartanracing.org/press/2.ht
ml?height=500&width=650&KeepThis=true. a

112 *"hunting, eating, and killing"* Joanne Pransky, "The Essential Interview: William
'Red' Whittaker, Field Robotics Pioneer and Entrepreneur," *Robotics Business Re-*

view, July 28, 2016, https://www.roboticsbusinessreview.com/research/essential-interview-william-red-whittaker-field-robotics-pioneer-entrepreneur/. a

113 *incubator for prematurely born infants* General Motors, "Cadillac's Electric Self Starter Turns 100," news release, February 12, 2015, https://media.gm.com/media/us/en/gm/home.detail.html/content/Pages/news/us/en/2012/Feb/0215_cad_starter.htm, a; "Charles F. Kettering," Lemelson-MIT Program, https://lemelson.mit.edu/resources/charles-f-kettering, accessed July 22, 2019; National Museum of the US Air Force, "Kettering Aerial Torpedo 'Bug,'" news release, April 7, 2015, https://www.nationalmuseum.af.mil/Visit/Museum-Exhibits/Fact-Sheets/Display/Article/198095/kettering-aerial-torpedo-bug/, accessed July 21, 2019.

113 *"at the same time as you do"* California Department of Motor Vehicles, *California Driver Handbook,* 2007, 16.

116 *keeping warbirds in the air* "Ecological Monitoring No Longer Needed at Castle," U.S. Air Force Civil Engineer Center, https://www.afcec.af.mil/Home/BRAC/Castle.aspx. a

116 *"Rabid bats are everywhere"* Michael Taylor, phone interview by author, Berkeley, California, January 6, 2019.

119 *a photo shoot for* WIRED Josh Davis, "Say Hello to Stanley," *WIRED,* January 1, 2006.

119 *shared their work in a series of research papers* Sebastian Thrun, Michael Montemerlo, Hendrik Dahlkamp, David Stavens, Andrei Aron, James Diebel, Philip Fong, John Gale, Morgan Halpenny, Kenny Lau, Celia Oakley, Mark Palatucci, Vaughan Pratt, Pascal Stang, Sven Strohband, Cedric Dupont, Lars-Erik Jendrossek, Christian Koelen, Charles Markey, Carlo Rummel, Joe Van Niekerk, Eric Jensen, Philippe Alessandrini, Gary Bradski, Bob Davies, Scott Ettinger, Adrian Kaehler, Ara Nefian, and Pamela Mahoney, "Stanley: The Robot That Won the DARPA Grand Challenge," *Journal of Field Robotics* 23, no. 9 (June 27, 2007); Sebastian Thrun, "A Personal Account of the Development of Stanley, the Robot That Won the DARPA Grand Challenge," *AI Magazine,* December 15, 2006, 69–82.

120 *faster way to scale that expansion* Bill Kilday, *Never Lost Again: The Google Mapping Revolution That Sparked New Industries and Augmented Our Reality* (New York: Harper Business, 2018).

121 *hit Page's million-mile milestone* Mark Harris, "God Is a Bot, and Anthony Levandowski Is His Messenger," *WIRED | Backchannel,* September 27, 2017, https://www.wired.com/story/god-is-a-bot-and-anthony-levandowski-is-his-messenger/. a

122 *twice what the 2005 Grand Challenge had cost Report to Congress: Darpa Prize Authority,* Defense Advanced Research Projects Agency, 2006; *Prizes for Advanced*

Technology Achievements: Fiscal Year 2007 Annual Report, Defense Advanced Research Projects Agency, January 2008.

124 *with a private plane full of executives* Lawrence D. Burns, *Autonomy: The Quest to Build the Driverless Car—And How It Will Reshape Our World* (New York: Ecco Press, 2018).

125 *"Not again"* Chris Urmson, in-person interview by author, Palo Alto, California, January 28, 2019.

127 *having completed its three missions* Elizabeth Svoboda, "DARPA Urban Challenge: We Have a . . . Finisher," *Popular Science*, November 3, 2007, https://www.popsci.com/article/2007-11/darpa-urban-challenge-we -have-finisher/.

128 *"That's what makes the great ones"* "Robotics History: Narratives and Networks Oral Histories: Red Whittaker," interview by Peter Asaro and Selma Šabanovic, IEEEtv, April 17, 2015, https://ieeetv.ieee.org/history/robotics-history-narra tives-and-networks-oral-histories-red-whittaker. a

128 *"I won that race"* William Whittaker, phone interview by author, San Francisco, May 13, 2017.

128 *"I didn't think it would happen in my lifetime"* Melanie Dumas, in-person interview by author, Mountain View, California, November 1, 2017.

Chapter 7

132 *Suzanna Musick, whose background* https://www.linkedin.com/in/suzannamusick/.

136 *building a car that could drive itself* Mark Harris, "God Is a Bot, and Anthony Levandowski Is His Messenger," *WIRED | Backchannel*, September 27, 2017, https://www.wired.com/story/god-is-a-bot-and-anthony-levandowski-is-his -messenger/, accessed July 28, 2019.

139 *"to the first ever autonomous pizza delivery vehicle"* "Prototype This! / Automated Pizza Delivery," Discovery, December 3, 2008.

140 *his past life as an academic* Sebastian Thrun, in-person interview by author, Mountain View, California, May 24, 2017.

140 *"It can't be done, goddammit"* Sebastian Thrun, in-person interview by author, Mountain View, California, February 1, 2019.

141 *bowing to short-term financial objectives* Larry Page and Sergey Brin, " 'An Owner's Manual' for Google's Shareholders," Alphabet Investor Relations, April 29, 2004, https://abc.xyz/investor/founders-letters/2004-ipo-letter/.

143 *"picking really hard roads"* Sebastian Thrun, in-person interview by author, Mountain View, California, May 24, 2017.

144 *"jogging start"* Chris Urmson, in-person interview by author, Palo Alto, California, January 28, 2019.

146 *"to win you World War II"* Isaac Taylor, in-person interview by author, Belmont, California, April 1, 2019.

146 *"He was damn right"* Sebastian Thrun, in-person interview by author, Mountain View, California, February 1, 2019.

147 *"not to do email every day"* Ibid.

151 *"We're going to make billions"* Anthony Levandowski, in-person interview by author, San Francisco, June 21, 2019.

152 *"until we got lucky"* Don Burnette, in-person interview by author, Mountain View, California, May 3, 2019.

153 *or Levandowski too aggressive* Charles Duhigg, "Did Uber Steal Google's Intellectual Property?," *New Yorker*, October 15, 2018.

154 *"a little bit of excitement in his day"* Chris Urmson, "How Google's Self-Driving Car Works," Lecture, 2011 IEEE/RSJ International Conference on Intelligent Robots and Systems, San Francisco, September 29, 2011, https://spectrum.ieee.org/automaton/robotics/artificial-intelligence/how-google-self-driving-car-works.

154 *the end of the Larry 1K* Burns, *Autonomy*.

155 *testing robot cars on public roads* John Markoff, *Machines of Loving Grace: The Quest for Common Ground Between Humans and Robots* (New York: Ecco Press, 2016).

155 *introduced the world to Google's self-driving car* John Markoff, "Google Cars Drive Themselves, in Traffic," *New York Times*, October 9, 2010, https://www.nytimes.com/2010/10/10/science/10google.html.

Chapter 8

161 *less likely to miss a pedestrian* Waymo, *Waymo Blog*, May 8, 2018, https://medium.com/waymo/google-i-o-recap-turning-self-driving-cars-from-science-fiction-into-reality-with-the-help-of-ai-89dded40c63.

162 *"aspects of this kind of technology"* "Minutes of the Meeting of the Assembly Committee on Transportation," April 7, 2011, https://www.leg.state.nv.us/Session/76th2011/Minutes/Assembly/TRN/Final/783.pdf.

163 *"they wouldn't come to our state"* Bruce Breslow, phone interview by author, Berkeley, California, April 23, 2019.

164 *signed his bill into law* Jerry Hirsch, "Brown Signs Bill Regulating Self-Driving Cars in California," *Los Angeles Times*, September 25, 2012, https://www.latimes.com/business/la-xpm-2012-sep-25-la-fi-mo-self-driving-car-law-20120925-story.html.

165 *Lawee told Thrun in an email* Uber opening statement, February 5, 2018, *Waymo LLC v. Uber Technologies, Inc.; Ottomotto LLC; Otto Trucking LLC* (United States District Court Northern District of California).

165 *"Anthony's commitment and integrity"* Ibid.

165 *murdering a couple along the way* Duhigg, "Did Uber Steal Google's Intellectual Property?"

165 *"He just changes the rules as he goes."* Seval Oz, phone interview by author, San Francisco, June 18, 2019.

165 *"team members will leave"* July 17, 2017, deposition of Larry Page, *Waymo LLC v. Uber Technologies, Inc.; Ottomotto LLC; Otto Trucking LLC* (United States District Court Northern District of California).

166 *would have gone to the employees.* See: Harris, Mark, "God Is a Bot, and Anthony Levandowski Is His Messenger." *WIRED | Backchannel*, September 27, 2017. Accessed July 28, 2019. https://www.wired.com/story/god-is-a-bot-and-anthony-levandowski-is-his-messenger/.

166 *"Hope I get to work with you again"* Kevin Rauwolf, in-person interview by author, Berkeley, California, April 2, 2019.

166 *"make Anthony rich if Chauffeur succeeds"* Uber opening statement, February 5, 2018, *Waymo LLC v. Uber Technologies, Inc.; Ottomotto LLC; Otto Trucking LLC* (United States District Court Northern District of California).

167 *commuting by car every day. A Look at Commuting Patterns in the United States from the American Community Survey*, US Census Bureau (2013).

168 *bonus from completing the Larry 1K* Burns *Autonomy.*

169 *"a complete waste of time"* Don Burnette, in-person interview by author, Mountain View, California, May 3, 2019.

169 *"disabled person to operate a car"* Burns, *Autonomy.*

170 *shared instead of privately owned* William J. Mitchell, Christopher E. Borroni-Bird, and Lawrence D. Burns, *Reinventing the Automobile: Personal Urban Mobility for the 21st Century* (Cambridge, MA: The MIT Press, 2010).

171 *laid out in Reinventing the Automobile* Burns, *Autonomy.*

171 *the German term that came to his mind,* über Adam Lashinsky, *Wild Ride: Inside Uber's Quest for World Domination* (New York: Portfolio/Penguin, 2017); Brad Stone, *The Upstarts: Uber, Airbnb, and the Battle for the New Silicon Valley* (New York: Back Bay Books/Little, Brown and Company, 2018). For subsequent references to Uber's origins and self-driving efforts, see same.

173 *"go the way of the DVD."* John Zimmer, *The Third Transportation Revolution: Lyft's Vision for the Next Ten Years and Beyond*, September 18, 2016, https://medium.com/@johnzimmer/the-third-transportation-revolution-27860f05fa91.

173 *"It was just not there yet."* Lashinsky, *Wild Ride.*

173 *"I can take the dude out of the front seat."* Stone, *Upstarts.*

174 *"because it was so much harder"* Sebastian Thrun, in-person interview by author, Mountain View, California, February 1, 2019.

175 *connected his phone to the computer* Chris Urmson, "How a Driverless Car Sees the Road," filmed March 2015 in Vancouver, Canada, TED video, 15:22, https://www.ted.com/talks/chris_urmson_how_a_driverless_car_sees_the _road?language=en.

175 *"We need to have this vehicle not need the driver"* Dave Ferguson, phone interview by author, Berkeley, California, April 22, 2019.

178 *auto industry supplier Roush to put it together* Aaron Foley, "Roush Is Helping Build Google's Self-Driving Car," Jalopnik, September 22, 2016, https://jalop nik.com/is-roush-building-googles-self-driving-car-1582788665.

178 *"be approachable and friendly"* Alex Davies, "Google's First Car: Revolutionary Tech in a Remarkably Lame Package," *WIRED*, May 28, 2014, https://www .wired.com/2014/05/google-self-driving-car-prototype/.

180 *it would be great news for riders* Kara Swisher, "The $17 Billion Man: Full Code Conference Video of Uber's Travis Kalanick," Recode, June 8, 2014, https:// www.vox.com/2014/6/8/11627734/the-17-billion-man-full-code-confer ence-video-of-ubers-travis-kalanick.

Chapter 9

184 *disengaged for a safety reason* Alex Davies, "Google's Self-Driving Cars Aren't as Good as Humans—Yet," *WIRED*, January 12, 2016, https://www.wired.com /2016/01/google-autonomous-vehicles-human-intervention/.

185 *"committed to making sure that doesn't happen"* Urmson, "How a Driverless Car Sees the Road."

185 *damaged by budget constraints* Uber opening statement, February 5, 2018, *Waymo LLC v. Uber Technologies, Inc.; Ottomotto LLC; Otto Trucking LLC* (United States District Court Northern District of California).

186 *"Let's make the right choice"* Ibid.

187 *the tension between Urmson and Levandowski* Burns, *Autonomy.*

188 *"that he could sell en masse to Uber"* Ryan Felton, " 'We Need to Fire Anthony': Google Before Finding Out Self-Driving Car Engineer Allegedly Stole Tech," Jalopnik, September 15, 2017, https://jalopnik.com/we-need-to-fire-anthony -google-before-finding-out-self-1811689572.

188 *" 'Anthony is doing that' "* July 17, 2017, deposition of Larry Page, *Waymo LLC v.*

Uber Technologies, Inc.; Ottomotto LLC; Otto Trucking LLC (United States District Court Northern District of California).

189 *"because he was already doing it"* Ibid.

190 *toward the Golden Gate Bridge* Mike Isaac, *Super Pumped: The Battle for Uber* (New York: W. W. Norton & Company, Inc., 2019).

190 *"brother from another mother."* February 7 testimony of Travis Kalanick, *Waymo LLC v. Uber Technologies, Inc.; Ottomotto LLC; Otto Trucking LLC* (United States District Court Northern District of California).

191 *"existential for Uber"* July 27, 2017, deposition of Travis Kalanick, *Waymo LLC v. Uber Technologies, Inc.; Ottomotto LLC; Otto Trucking LLC* (United States District Court Northern District of California).

191 *"deploy 100,000 cars in 2020"* Waymo opening statement, February 5, 2018, *Waymo LLC v. Uber Technologies, Inc.; Ottomotto LLC; Otto Trucking LLC* (United States District Court Northern District of California).

191 *"Laser is the sauce"* Testimony of Travis Kalanick, February 6, 2018, *Waymo LLC v. Uber Technologies, Inc.; Ottomotto LLC; Otto Trucking LLC* (United States District Court Northern District of California).

191 *"Uber had to ship a product"* Anthony Levandowski, in-person interview by the author, San Francisco, June 21, 2019.

191 *"Chauffeur is broken"* Testimony of John Krafcik, February 5, 2018, *Waymo LLC v. Uber Technologies, Inc.; Ottomotto LLC; Otto Trucking LLC* (United States District Court Northern District of California).

192 *"We have a tentative deal with respect to Newco"* July 27, 2017, deposition of Travis Kalanick, *Waymo LLC v. Uber Technologies, Inc.; Ottomotto LLC; Otto Trucking LLC* (United States District Court Northern District of California).

192 *"there's just too much BS"* July 17, 2017, deposition of Larry Page, *Waymo LLC v. Uber Technologies, Inc.; Ottomotto LLC; Otto Trucking LLC* (United States District Court Northern District of California).

193 *$120 million for the engineer* Mark Harris, "God Is a Bot, and Anthony Levandowski Is His Messenger," *WIRED | Backchannel*, September 27, 2017, https://www.wired.com/story/god-is-a-bot-and-anthony-levandowski-is-his-messenger/, accessed July 28, 2019.

193 *"like I'm in the trunk"* Charles Duhigg, "Did Uber Steal Google's Intellectual Property?," *New Yorker*, October 15, 2018.

193 *latest work of the team's technical leaders* Declaration of Gary Brown, signed March 9, 2017, *Waymo LLC v. Uber Technologies, Inc.; Ottomotto LLC; Otto Trucking LLC* (United States District Court Northern District of California).

194 *Google would "crush" him* "Summary Report: Project Unicorn Investigation" by

Stroz Friedberg, filed October 2, 2017, *Waymo LLC v. Uber Technologies, Inc.; Ottomotto LLC; Otto Trucking LLC* (United States District Court Northern District of California).

194 *they wrote in a blog post* Otto, "Introducing Otto, the Startup Rethinking Commercial Trucking," Official Otto Blog, May 17, 2016, https://blog.ot.to/introducing-otto-the-startup-rethinking-commercial -trucking-cfdc502ef452.

194 *any penalties for violating its rules* Mark Harris, "How Otto Defied Nevada and Scored a $680 Million Payout from Uber," *WIRED*, January 8, 2018, https://www.wired.com /2016/11/how-otto-defied-nevada-and-scored-a-680-million-payout-from-uber/.

195 *"pushing for the right things"* Anthony Levandowski text to Travis Kalanick, filed August 11, 2017, *Waymo LLC v. Uber Technologies, Inc.; Ottomotto LLC; Otto Trucking LLC* (United States District Court Northern District of California).

196 *announcing his departure* Chris Urmson, "The View from the Front Seat of the Google Self-Driving Car: A New Chapter," Medium, August 5, 2016. https:// medium.com/@chris_urmson/the-view-from-the-front-seat-of-the-google -self-driving-car-a-new-chapter-7060e89cb65f#.wfldpz85w.

196 *"we should be seriously concerned"* Uber opening statement, February 5, 2018, *Waymo LLC v. Uber Technologies, Inc.; Ottomotto LLC; Otto Trucking LLC* (United States District Court Northern District of California).

196 *"F-you money"* Alistair Barr and Mark Bergen, "One Reason Staffers Quit Google's Car Project? The Company Paid Them So Much," *Bloomberg*, February 13, 2017, https://www.bloomberg.com/news/articles/2017-02-13/one -reason-staffers-quit-google-s-car-project-the-company-paid-them-so-much.

197 *"I think there should be no rules"* Anthony Levandowski, phone interview by author, Berkeley, California, March 4, 2018.

198 *revoking the registrations for the sixteen test cars* Aarian Marshall, "Uber Bows Before California's Power and Parks Its Robo-Cars," *WIRED*, December 21, 2016, https://www.wired.com/2016/12/uber-parks-its-robo-cars/.

Chapter 10

200 *"going from the horse to the car"* John Casesa, phone interview by author. San Francisco, May 15, 2018.

200 *on the road to bankruptcy* Bryce G. Hoffman, *American Icon: Alan Mulally and the Fight to Save Ford Motor Company* (New York: Crown, 2013). For subsequent references to Ford between 2006 and 2013, see same.

201 *especially not in urban areas* Gary Silberg, "The Clockspeed Dilemma," KPMG, November 2015, https://assets.kpmg/content/dam/kpmg/pdf/2016/04/auto

-clockspeed-dilemma.pdf; Jerry Hirsch, "Major Auto Industry Disruption Will Lead to Robotic Taxis, Morgan Stanley Says," *Los Angeles Times,* April 7, 2015, https://www .latimes.com/business/autos/la-fi-hy-end-of-human-driving-20150407-story.html.

201 *"And it won't stop."* Adam Jonas, "Shared Autonomy: Put This Chart On Your Wall, It's My Sad Life," Morgan Stanley, April 7, 2015, https://orfe.princeton.edu /~alaink/SmartDrivingCars/PDFs/MorganStanley%20040715ReportJonas.pdf.

201 *another $500 million in 2018* Alex Wilhelm, "A Quick Peek at Zoox's $500M Round," *Crunchbase News,* July 18, 2018, https://news.crunchbase.com/news /a-quick-peek-at-zooxs-500m-round/.

202 *"new opportunities leading to profitable growth"* "Ford Hires John Casesa to Lead Global Strategy," Ford Media Center, Ford Motor Company, February 17, 2015, https://media.ford.com/content/fordmedia/fna/us/en/news/2015/02 /17/executive-announcement.html.

202 *"on the bellows, cranking away"* Edwin Olson, phone interview by author, San Francisco, January 31, 2019.

203 *"help change the way the world moves"* "Ford at CES Announces Smart Mobility Plan and 25 Global Experiments Designed to Change the Way the World Moves," Ford Media Center, Ford Motor Company, January 6, 2015, https:// media.ford.com/content/fordmedia/fna/us/en/news/2015/01/06/ford-at -ces-announces-smart-mobility-plan.html.

204 *could make their cars more compelling* Alex Davies, "Ford Finally Discovers Silicon Valley," *WIRED,* January 23, 2015, https://www.wired.com/2015/01/ ford-silicon-valley-research-center/.

204 *ceded control of a major part of the in-car experience to Google and Apple* Derek Viita, "Consumer Interest for In-Car Smartphone Mirroring Is Almost Universal," *Strategy Analytics,* October 12, 2017, https://www.strategyanalytics.com /access-services/automotive/in-vehicle-ux/reports/report-detail/consumer -interest-for-in-car-smartphone-mirroring-is-almost-universal#.Wd942Wh Sw2w0.3067323840854923.

204 *infotainment mattered more than horsepower* Tom Anderson, "Commuters Waste a Full Week in Traffic Each Year," CNBC, August 10, 2016, https://www.cnbc .com/2016/08/09/commuters-waste-a-full-week-in-traffic-each-year.html.

206 *"donkey ride through hell"* Tory Smith, in-person interview by author, San Francisco, November 7, 2018.

207 *In early January of 2016, it pulled away* Sharon Silke Carty, "Failed Google Deal Left Fields in the Lurch," *Automotive News,* May 29, 2017, https://www.au tonews.com/article/20170529/OEM/170529795/failed-google-deal-left -fields-in-the-lurch; Burns, *Autonomy.*

207 *exactly the sort of story Ford had hoped for* Bill E. Vlasic and Neal E. Boudette, "Google to Get Fiat Chrysler Minivans for Self-Driving Tests," *New York Times*, May 3, 2016, https://www.nytimes.com/2016/05/04/technology/google-fiat-chrysler-minivans-self-driving.html.

208 *"We're totally bought in on that"* Alisyn Malek, in-person interview by author, South San Francisco, February 5, 2019.

209 *"How is that a job?"* Ibid.

209 "get around to shooting at some point." Byron Shaw, phone interview by author, Berkeley, California, January 22, 2019.

210 *"a Versailles of Industry"* "Architecture for the Future: GM Constructs a 'Versailles of Industry,'" *Life*, May 21, 1956.

211 *Burns had declined* Burns, *Autonomy*.

214 *"provide them with choices that make their life better and easier"* Mary Barra, "The Year Detroit Takes on Silicon Valley," LinkedIn, December 14, 2015, https://www.linkedin.com/pulse/big-idea-2016-year-detroit-takes-silicon-valley-mary-barra/.

214 *jointly create a self-driving ridehail service* Alex Davies, "GM and Lyft Are Building a Network of Self-Driving Cars," *WIRED*, January 4, 2016, https://www.wired.com/2016/01/gm-and-lyft-are-building-a-network-of-self-driving-cars/.

214 *"one day I'll be able to fly"* Urmson, "How a Driverless Car Sees the Road."

216 *sent them a congratulatory tweet* Alex Davies, "Obviously Drivers Are Already Abusing Tesla's Autopilot," *WIRED*, October 22, 2015, https://www.wired.com/2015/10/obviously-drivers-are-already-abusing-teslas-autopilot/.

216 *for designing a system that allowed that inattention* "Collision Between a Car Operating with Automated Vehicle Control Systems and a Tractor-Semitrailer Truck Near Williston, Florida, May 7, 2016," National Transportation Safety Board (NTSB), September 12, 2017, https://www.ntsb.gov/investigations/AccidentReports/Reports/HAR1702.pdf.

217 *"like something a robot could do"* Kyle Vogt, in-person interview by author, San Francisco, July 13, 2016.

217 *"emerge with something that worked"* Erin Griffith, "Driven in the Valley: The Startup Founders Fueling GM's Future," *Fortune*, September 22, 2016, https://fortune.com/longform/cruise-automation-general-motors-driverless-cars/.

217 *sold Twitch to Amazon for $970 million* Griffith, "Driven in the Valley."

218 *" 'OK, everything's under control.' "* Dan Ammann, in-person interview by author, San Francisco, May 29, 2019.

218 *"That's pretty interesting"* Ibid.

219 *he meant hundreds of vehicles* Alex Davies, "Ford Says It'll Have a Fleet of Fully Autonomous Cars in Just 5 Years," *WIRED*, August 16, 2016, https://www.wired.com/2016/08/ford-autonomous-vehicles-2021/.

220 *"centered around the automation of the automobile"* Mark Fields, remarks to press, Palo Alto, California, August 16, 2016.

220 *"create a software company"* John Casesa, phone interview by author, San Francisco, May 15, 2018.

222 *"I wouldn't bet against him"* Anthony Levandowski, in-person interview by author, San Francisco, June 21, 2019.

Chapter 11

223 *resembled that of an "explorer's ship"* "Aerial View of Carnegie Technical Institute Campus," Historic Pittsburgh, University of Pittsburgh Library System, https://historicpittsburgh.org/islandora/object/pitt:MSP285.B004.F02.I05, accessed September 7, 2019.

223 *hardware that students now created there* "Undergraduate Walking Tour: History of Hamerschlag Hall," Carnegie Mellon University, https://www.archive.ece .cmu.edu/about/_files/walking_tour_guide_web.pdf, accessed September 6, 2019.

227 *"You're off the hook!"* Evan Ackerman, "Carnegie Mellon Solves 12-Year-Old DARPA Grand Challenge Mystery," *IEEE Spectrum*, October 19, 2017, https:// spectrum.ieee.org/cars-that-think/transportation/self-driving/cmu-solves-12 -year-old-darpa-grand-challenge-mystery.

228 *for whoever could make that happen* Roger Lanctot, "Intel Predicts Autonomous Driving Will Spur New 'Passenger Economy' Worth $7 Trillion," *Strategy Analytics*, June 2017, https://newsroom.intel.com/news-releases/intel-predicts-auton omous-driving-will-spur-new-passenger-economy-worth-7-trillion/#gs.o1eivj.

228 *it was now called Waymo* Alex Davies, "Google's Self-Driving Car Company Is Finally Here," *WIRED*, December 13, 2016, https://www.wired.com/2016/12 /google-self-driving-car-waymo/.

229 *"ethical behavior that's out there"* Chris Urmson, phone interview by author, Pelham, New York, June 5, 2017.

229 *could land Levandowski in prison* Aarian Marshall, "Google's Fight Against Uber Takes a Turn for the Criminal," *WIRED*, May 12, 2017, https://www.wired .com/2017/05/googles-fight-uber-takes-turn-criminal/.

230 *"anybody doing something that bad"* Ibid.

230 *had destroyed the discs containing them* Stroz Friedberg, "Summary Report: Project Unicorn Investigation," filed October 2, 2017, *Waymo LLC v. Uber Technologies, Inc.; Ottomotto LLC; Otto Trucking LLC* (United States District Court Northern District of California).

231 *presented the Otto acquisition to his board of directors* Klint Finley, "One of Uber's

First Investors Sued Travis Kalanick for Fraud," *WIRED*, August 10, 2017, https://www.wired.com/story/benchmark-capital-just-sued-former-uber-ceo -travis-kalanick-for-fraud/.

232 *any of Waymo's hardware or software in its self-driving cars* Aarian Marshall, "Uber and Waymo Abruptly Settle for $245 Million," *WIRED*, February 9, 2018, https://www.wired.com/story/uber-waymo-lawsuit-settlement/.

232 *"This case is ancient history"* Judicial proceedings on February 9, 2018, *Waymo LLC v. Uber Technologies, Inc.; Ottomotto LLC; Otto Trucking LLC* (United States District Court Northern District of California).

232 *machines were bound to rule over humans.* Mark Harris, "God Is a Bot, and Anthony Levandowski Is His Messenger," *WIRED | Backchannel*, September 27, 2017, accessed July 28, 2019. https://www.wired.com/story/god-is-a-bot-and -anthony-levandowski-is-his-messenger/.

232 *"We're basically creating God."* Anthony Levandowski, in-person interview by author, San Francisco, June 21, 2019.

232 *encounter with the Toyota Camry* Charles Duhigg, "Did Uber Steal Google's Intellectual Property?" *New Yorker*, October 15, 2018.

233 *"a work of fiction"* Mark Harris, "Anthony Levandowski Faces New Claims of Stealing Trade Secrets," *WIRED*, January 16, 2018, https://www.wired.com/ story/anthony-levandowski-faces-new-claims-of-stealing-trade-secrets/.

233 *carried the headline.* Justin T. Westbrook, "The Engineer in the Google vs. Uber 'Stolen Tech' Case Really Was Terrible," Jalopnik, October 16, 2018, https://jalopnik.com/the-engineer-in-the-google-vs-uber-stolen-tech-case -1829768721.

233 *"afraid to have him around"* Dave Hall, in-person interview by author, Alameda, California, November 18, 2018.

233 *"anywhere near my ethical standards."* Sebastian Thrun, in-person interview by author, Mountain View, California, February 1, 2019.

233 *"a weasel"* Luc Vincent, in-person interview by author, San Francisco, February 14, 2019.

233 *"evil"* Kyle Machulis, in-person interview by author, Berkeley, California, March 6, 2019.

234 *its suite of sensors and self-driving software* "Accident Report Detail," National Transportation Safety Board (NTSB), May 24, 2018, https://www.ntsb. gov/investigations/AccidentReports/Pages/HWY18MH010-prelim.aspx. For subsequent references to details of Uber crash, see same.

236 *would not have made the same mistake* Faiz Siddiqui, "Waymo's CEO on Fatal Autonomous Uber Crash: Our Car Would Have Been Able to Handle It,"

Washington Post, March 25, 2018, https://www.washingtonpost.com/local/ trafficandcommuting/waymos-ceo-on-fatal-autonomous-uber-crash-our-car -would-have-been-able-to-handle-it/2018/03/25/4cc97550-3046-11e8-8abc -22a366b72f2d_story.html.

236 *later published by* The Information's *Amir Efrati,* "The Uber Whistleblower's Email," *The Information,* December 10, 2018, https://www.theinformation .com/articles/the-uber-whistleblowers-email.

236 *would look into his concerns* Robbie Miller, Signal message exchange with author, Berkeley, California, November 11, 2019.

237 *pleaded not guilty* Kate Conger, "Driver Charged in Uber's Fatal 2018 Autonomous Car Crash," *New York Times,* September 15, 2020, https://www.nytimes .com/2020/09/15/technology/uber-autonomous-crash-driver-charged.html.

237 *it had set her up to fail* Alex Davies, "The Unavoidable Folly of Making Humans Train Self-Driving Cars," *WIRED,* June 22, 2018, https://www.wired.com/ story/uber-crash-arizona-human-train-self-driving-cars/.

237 *"with open arms and wide open roads"* "Governor Ducey Tells Uber 'CA May Not Want You, But AZ Does,'" Office of the Arizona Governor, December 22, 2016, https://azgovernor.gov/governor/news/2016/12/governor-ducey-tells-uber -ca-may-not-want-you-az-does.

238 *and pull over safely* Aarian Marshall, "32 Hours in Chandler, Arizona, the Self-Driving Capital of the World," *WIRED,* September 2018, https://www.wired .com/story/32-hours-chandler-arizona-self-driving-capital/.

239 *had nearly hit their ten-year-old son.* Simon Romero, "A Growing Test for Self-Driving Vehicles: Rocks, Knives and Guns," *New York Times,* December 31, 2018.

240 *"world's most experienced driver"* "Waymo," https://waymo.com/, accessed September 23, 2019.

Epilogue

244 *the car going under the speed limit* Mark Harris, "Self-Driving Car Drove Me from California to New York, Claims Ex-Uber Engineer," *Guardian,* December 18, 2018, https://www.theguardian.com/technology/2018/dec/18/controversial -engineer-i-travelled-over-3000-miles-in-a-self-driving-car.

244 *his latest company, Pronto* Anthony Levandowski, "Pronto Means Ready," *Medium,* Pronto AI, December 21, 2018, https://medium.com/pronto-ai/pronto -means-ready-e885bc8ec9e9.

246 *"I'm trying to get onto the garage roof"* Anthony Levandowski, in-person interview by author, San Francisco, May 16, 2019.

246 *and attempted theft of trade secrets* "Former Uber Self-Driving Car Executive In-
     dicted for Alleged Theft of Trade Secrets from Google," United States Depart-
     ment of Justice, August 27, 2019, https://www.justice.gov/usao-ndca/pr/former
     -uber-self-driving-car-executive-indicted-alleged-theft-trade-secrets-google.

247 *filed for bankruptcy* Reed Albergotti, "A former Uber executive was ordered to
     pay Google $179 million. Then he filed for bankruptcy," *The Washington Post*,
     March 4, 2020, https://www.washingtonpost.com/technology/2020/03/04
     /levandowski-bankrupt/.

247 *"I'm happy to put this behind me"* Anthony Levandowski, phone interview by
     author, Berkeley, California, March 19, 2020.

247 *to set an example.* Judicial proceedings on August 4, 2020, *United States of
     America, Plaintiff, v. Anthony Scott Levandowski, Defendant* (United States
     District Court Northern District of California).

248 *"Why I Went to Federal Prison."* Judgment in a criminal case, August 6, 2020,
     United States of America, Plaintiff, v. Anthony Scott Levandowski, Defendant
     (United States District Court Northern District of California).

248 *"disabled person to operate a car"* Burns, *Autonomy*.

248 *didn't offer a revised timeline* Alex Davies, "GM's Cruise Rolls Back Its Target for
     Self-Driving Cars," *WIRED*, July 24, 2019, https://www.wired.com/story/gms
     -cruise-rolls-back-target-self-driving-cars/.

248 *costing the economy more than $300 billion* "Los Angeles Tops INRIX Global
     Congestion Ranking," Inrix, February 5, 2018, http://inrix.com/press-releases
     /scorecard-2017/.

248 *"out in the middle of nowhere."* Sal Fish. Phone interview by author. San Fran-
     cisco, May 18, 2017.

248 *boosting the technology that would come too late to save him* Karen Jonas, "Slash
     -X Owner Killed in Head-on Collision," *Daily Press*, August 12, 2011, jhttps://
     www.vvdailypress.com/article/20110812/NEWS/308129982.

249 *"peak of inflated expectations"* "Hype Cycle Research Methodology," *Gartner*,
     https://www.gartner.com/en/research/methodologies/gartner-hype-cycle,
     accessed September 12, 2019.

250 *"you can solve with fifty or a hundred engineers"* Dan Amman, in-person interview
     by author, San Francisco, May 29, 2019.

250 *"with as few miles as possible"* Don Burnette, in-person interview by author,
     Mountain View, California, May 3, 2019.

252 *"graphic solution to a problem which they all faced"* Norman Bel Geddes, *Magic
     Motorways* (New York: Random House, 1940).

Index